AF326848

BIOLOGICAL AND MEDICAL PHYSICS,
BIOMEDICAL ENGINEERING

BIOLOGICAL AND MEDICAL PHYSICS, BIOMEDICAL ENGINEERING

The fields of biological and medical physics and biomedical engineering are broad, multidisciplinary and dynamic. They lie at the crossroads of frontier research in physics, biology, chemistry, and medicine. The Biological and Medical Physics, Biomedical Engineering Series is intended to be comprehensive, covering a broad range of topics important to the study of the physical, chemical and biological sciences. Its goal is to provide scientists and engineers with textbooks, monographs, and reference works to address the growing need for information.

Books in the series emphasize established and emergent areas of science including molecular, membrane, and mathematical biophysics; photosynthetic energy harvesting and conversion; information processing; physical principles of genetics; sensory communications; automata networks, neural networks, and cellular automata. Equally important will be coverage of applied aspects of biological and medical physics and biomedical engineering such as molecular electronic components and devices, biosensors, medicine, imaging, physical principles of renewable energy production, advanced prostheses, and environmental control and engineering.

Y. Takeuchi   Y. Iwasa   K. Sato   (Eds.)

Mathematics for Life Science and Medicine

With 31 Figures

Prof. Yasuhiro Takeuchi
Shizuoka University
Faculty of Engineering
Department of Systems Engineering
Hamamatsu 3-5-1
432-8561 Shizuoka
Japan
email: takeuchi@sys.eng.shizuoka.ac.jp

Prof. Yoh Iwasa
Kyushu University
Department of Biology
812-8581 Fukuoka
Japan
e-mail: yiwasscb@mbox.nc.kyushu-u.ac.jp

Dr. Kazunori Sato
Shizuoka University
Faculty of Engineering
Department of Systems Engineering
Hamamatsu 3-5-1
432-8561 Shizuoka
Japan
email: sato@sys.eng.shizuoka.ac.jp

Library of Congress Cataloging in Publication Data: 2006931400

ISSN 1618-7210

ISBN-10 3-540-34425-X Springer Berlin Heidelberg New York

ISBN-13 978-3-540-34425-4 Springer Berlin Heidelberg New York

Springer is a part of Springer Science+Business Media

springer.com

© Springer-Verlag Berlin Heidelberg 2007

Cover concept by eStudio Calamar Steinen

Cover production: WMXDesign GmbH, Heidelberg
Production: LE-TeX Jelonek, Schmidt, Vöckler GbR, Leipzig

Printed on acid-free paper   SPIN 10995921   57/3141/NN - 5 4 3 2 1 0

Preface

Dynamical systems theory in mathematical biology and environmental science has attracted much attention from many scientific fields as well as mathematics. For example, "chaos" is one of its typical topics. Recently the preservation of endangered species has become one of the most important issues in biology and environmental science, because of the recent rapid loss of biodiversity in the world. In this respect, permanence and persistence, the new concepts in dynamical systems theory, are important. These give a new aspect in mathematics that includes various nonlinear phenomena such as chaos and phase transition, as well as the traditional concepts of stability and oscillation. Permanence and persistence analyses are expected not only to develop as new fields in mathematics but also to provide useful measures of robust survival for biological species in conservation biology and ecosystem management. Thus the study of dynamical systems will hopefully lead us to a useful policy for bio-diversity problems and the conservation of endangered species. This brings us to recognize the importance of collaborations among mathematicians, biologists, environmental scientists and many related scientists as well. Mathematicians should establish a mathematical basis describing the various problems that appear in the dynamical systems of biology, and feed back their work to biology and environmental sciences. Biologists and environmental scientists should clarify/build the model systems that are important in their own as global biological and environmental problems. In the end mathematics, biology and environmental sciences develop together.

The International Symposium "Dynamical Systems Theory and Its Applications to Biology and Environmental Sciences", held at Hamamatsu, Japan, March 14th-17th, 2004, under the chairmanship of one of the editors (Y.T.), gave the editors the idea for the book *Mathematics for Life Science and Medicine* and the chapters include material presented at the symposium as invited lectures.

The editors asked authors of each chapter to follow some guidelines:

1. to keep in mind that each chapter will be read by many non-experts, who do not have background knowledges of the field;
2. at the beginning of each chapter, to explain the biological background of the modeling and theoretical work. This need not include detailed information about the biology, but enough knowledge to understand the model in question;
3. to review and summarize the previous theoretical and mathematical works and explain the context in which their own work is placed;
4. to explain the meaning of each term in the mathematical models, and the reason why the particular functional form is chosen, what is different from other authors' choices etc. What is obvious for the author may not be obvious for general readers;
5. then to present the mathematical analysis, which can be the main part of each chapter. If it is too technical, only the results and the main points of the technique of the mathematical analysis should be presented, rather than showing all the steps of mathematical proof;
6. at the end of each chapter, to have a section ("Discussion") in which the author discusses biological implications of the outcome of the mathematical analysis (in addition to mathematical discussion).

Mathematics for Life Science and Medicine includes a wide variety of stimulating fields, such as epidemiology, and gives an overview of the theoretical study of infectious disease dynamics and evolution. We hope that the book will be useful as a source of future research projects on various aspects of infectious disease dynamics. It is also hoped that the book will be useful to graduate students in the mathematical and biological sciences, as well as to those in some areas of engineering and medicine. Readers should have had a course in calculus, and knowledge of basic differential equations would be helpful.

We are especially pleased to acknowledge with gratitude the sponsorship and cooperation of Ministry of Education, Sports, Science and Technology, The Japanese Society for Mathematical Biology, The Society of Population Ecology, Mathematical Society of Japan, Japan Society for Industrial and Applied Mathematics, The Society for the Study of Species Biology, The Ecological Society of Japan, Society of Evolutionary Studies, Japan, Hamamatsu City and Shizuoka University, jointly with its Faculty of Engineering; Department of Systems Engineering.

Special thanks should also go to Keita Ashizawa for expert assistance with TEX. Drs. Claus Ascheron and Angela Lahee, the editorial staff of Springer-Verlag in Heidelberg, are warmly thanked.

Shizouka, *Yasuhiro Takeuchi*
Fukuoka, *Yoh Iwasa*
June 2006 *Kazunori Sato*

Contents

List of Contributors

Edoardo Beretta
Institute of Biomathematics,
University of Urbino,
Italy
e.beretta@mat.uniurb.it

Margherita Carletti
Biomathematics,
University of Urbino,
Italy
m.carletti@mat.uniurb.it

H.I. Freedman
Department of Mathematical,
and Statistical Sciences,
University of Alberta,
Edmonton, Alberta,
Canada
hfreedma@math.ualberta.ca

Yoh Iwasa
Department of Biology,
Faculty of Sciences, Kyushu
University,
Japan
yiwasscb@mbox.nc.kyushu-u.ac.jp

Denise E. Kirschner
Dept. of Microbiology and Immunol-
ogy,
University of Michigan Medical
School,
USA
kirschne@umich.edu

Jianquan Li
Department of Mathematics and
Physics,
Air Force Engineering University,
China
jianq_li@263.net

Wanbiao Ma
Department of Mathematics and
Mechanics,
School of Applied Science,
University of Science and Technology
Beijing,
China
wanbiao_ma@sas.ustb.edu.cn

Zhien Ma
Department of Applied Mathemat-
ics,
Xi'an Jiaotong University,
China
zhma@mail.xjtu.edu.cn

Simeone Marino
Dept. of Microbiology and Immunol-
ogy,
University of Michigan Medical
School,
USA
simeonem@umich.edu

Franziska Michor
Program in Evolutionary Dynamics,
Harvard University,
USA

Martin Nowak
Program in Evolutionary Dynamics,
Harvard University,
USA
nowakmar@omega.im.wsp.zgora.pl

Shigui Ruan
Department of Mathematics,
University of Miami,
USA
ruan@math.miami.edu

Kazunori Sato
Department of Systems Engineering,
Faculty of Engineering,
Shizuoka University,
Japan
sato@sys.eng.shizuoka.ac.jp

Yasuhiro Takeuchi
Department of Systems Engineering,
Faculty of Engineering,
Shizuoka University,
Japan
takeuchi@sys.eng.shizuoka.ac.jp

Horst R. Thieme
Department of Mathematics and
Statistics,
Arizona State University,
U.S.A.
h.thieme@asu.edu

Wendi Wang
Department of Mathematics,
Southwest China Normal University,
China
wendi@swnu.edu.cn

1

Mathematical Studies of Dynamics and Evolution of Infectious Diseases

Yoh Iwasa, Kazunori Sato, and Yasuhiro Takeuchi

The practical importance of understanding the dynamics and evolution of infectious diseases is steadily increasing in the contemporary world. One of the most important mortality factors for the human population is malaria. Every year, hundreds of millions of people suffer from malaria, and more than a million children die. One of the obstacles of controlling malaria is the emergence of drug-resistant strains. Pathogen strains resistant to antibiotics pose an important threat in developing countries. In addition, we observe new infectious diseases, such as HIV, Ebora, and SARS.

The mathematical study of infectious disease dynamics has a long history. The classic work by Kermack and McKendrick (1927) established the basis of modeling infectious disease dynamics. The variables indicate the numbers of host individuals in several different states – susceptive, infective and removed. This formalism is the basis of all current modeling of the dynamics and evolution of infectious diseases. Since then, the number of theoretical papers on infectious diseases has increased steadily. Especially influential was a series of papers by Roy Anderson and Robert May, summarized in their book (Anderson and May 1991). Anderson and May have developed population dynamic models of the host engaged in reproduction and migration. In a sense, they treated epidemic dynamics as a variant of ecological population dynamics of multiple species community. Combining the increase of our knowledge of nonlinear dynamical systems (e. g. chaos), Anderson and May also demonstrated the usefulness of simple models in understanding the basic principles of the system, and sometimes even in choosing a proper policy of infectious disease control.

The dynamical systems for epidemics are characterized by nonlinearity. The systems include many processes at very different scales, from the population on earth to the individual level, and further to the immune system within a patient. Hence, mathematical studies of epidemics need to face this dynamical diversity of phenomena. Tools of modeling and analysis for situations including time delay and spatial heterogeneity are very important. As a consequence, there is no universal mathematical model that holds for all problems in epidemics. When we are given a set of epidemiological phenomena and questions to answer, we must "construct" mathematical models that can describe the phenomena and answer our questions. This is quite different from studies in "pure" mathematics, in which usually the models are given beforehand.

One of the most important questions in mathematical studies of epidemics is the possibility of the eradication of disease. The standard local stability analysis of the endemic equilibrium and disease-free equilibrium is often not enough to answer the question, because it gives us information only on the local behavior, or the solution in the neighborhood of those equilibria. On the other hand, it is known that global stability analysis of the models is often very difficult, and even impossible in general cases, because the dynamics are highly nonlinear. Even if the endemic equilibrium were unstable and the disease-free equilibrium were locally stable, the diseases can remain endemic and be sustained forever. Sometimes, rather simple models show periodic or chaotic behavior. Recently, the concept of "permanence" was introduced in population biology and has been studied extensively. This concept is very important in mathematical epidemiology as well. Permanence implies that the disease will be maintained globally, irrespective of the initial composition. Even if the endemic equilibrium were unstable, the disease will last forever, possibly with perpetual oscillation or chaotic fluctuation.

Since the epidemiological data supplied by medical and public health sectors are abundant, epidemiological models are in general much better tested than similar population models in ecology developed for wild animals and plants. The diversity of models is also extensive, including all the different levels of complexity. Rather simple and abstract models are suitable to discuss general properties of the system, while more complex and realistic computer-based simulators are adopted for policy decision making incorporating details of the structure closely corresponding to available data. Mathematical modeling of infectious diseases is the most advanced subfield of theoretical studies in biology and the life sciences. What is notable in this development is that, even if many computer-based detailed simulators become available, the rigorous mathematical analysis of simple models remains very useful, medically and biologically, in giving a clear understanding of the behavior of the system.

Recently, the evolutionary change of infectious agents in the host population or within a patient has attracted an increasing attention. Mutations during genome replication would create pathogens that may differ slightly from the original types. This gives an opportunity for a novel strain to emerge and spread. As noted before, emergence of resistant strains is a major obstacle of infectious disease control. Essentially the same evolutionary process occurs within the body of a single patient. A famous example is HIV, in which viral particles change and diversify their nucleotide sequences after they infect a patient. This supposedly reflects the selection by the immune system of the host working on the virus genome. A similar process of escape is involved in carcinogenesis – a process in which normal stem cells of the host become cancerous.

The papers included in this volume are for mathematical studies of models on infectious diseases and cancer. Most of them are based on presentations in the First International Symposium on Dynamical Systems Theory and its Applications to Biology and Environmental Sciences, held in Hamamatsu, Japan, on 14–17 March 2004. This introductory chapter is followed by four papers on infectious disease dynamics, in which the roles of time delay (Chaps. 2 and 3) and spatial structures (Chaps. 4 and 5) are explored. Then, there are two chapters that discuss competition between strains and evolution occurring in the host population (Chap. 6) and within a single patient (Chap. 7). Finally, there are papers on models of the immune system and cancer (Chaps. 8 and 9). Below, we briefly summarize the contents of each chapter.

In Chap. 2, Zhien Ma and Jianquan Li give an introduction to the mathematical modeling of disease dynamics. Then, they summarize a project of modeling the spread of SARS in China by the authors and their colleagues.

In Chap. 3, Yasuhiro Takeuchi and Wanbiao Ma introduce mathematical studies of models with time delay. They first review past mathematical studies on this theme during the last few decades, and then introduce their own work on the stability of the equilibrium and the permanence of epidemiological dynamics.

In Chaps 4 and 5, Wendi Wang and Shigui Ruan discuss the spatial aspect of epidemiology. The spread of a disease in a population previously not infected may appear as "wave of advance". This is often modeled as a reaction diffusion system, or by other models handling spatial aspects of population dynamics. The speed of disease propagation is analogous to the spread of invaders in a novel habitat in spatial ecology (Shigesada and Kawasaki 1997).

Since microbes have a shorter generation time and huge numbers of individuals, they have much faster evolutionary changes, causing drug resistance and immune escape, among the most common problems in epidemiology. By considering the appearance of novel strains with different properties from those of the resident population of pathogens, and tracing their abundance, we can discuss the evolutionary dynamics of infectious diseases. In Chap. 6, Horst Thieme summarized the work on the competition between different and competing strains, and the possibility of their coexistence and replacement. An important concept is the "maximal basic replacement ratio". If a host once infected and then recovered from a single strain is perfectly immune to all the other strains (i. e. cross immunity is perfect), then the one with the largest basic replacement ratio will win the competition among the strains. The author explores the extent to which this result can be generalized. He also discusses the coexistence of strains considering the aspect of maternal transmission as well.

In Chap. 7, Yoh Iwasa and his colleagues analyze the result of evolutionary change occurring within the body of a single patient. Some of the pathogens, especially RNA viruses have high mutation rates, due to an unreliable replication mechanism, and hence show rapid genetic change in a host. The nucleotide sequences just after infection by HIV will be quite different from those HIV occurring after several years. By mutation and natural selection under the control of the immune system, they become diversified and constantly evolve. Iwasa and his colleagues derive a result that, without cross-immunity among strains, the pathogenicity of the disease tends to increase by any evolutionary changes. They explore several different forms of cross-immunity for which the result still seems to hold.

In Chap. 8, Edoardo Beretta and his colleagues discuss immune response based on mathematical models including time delay. The immune system has evolved to cope with infectious diseases and cancers. They have properties of immune memory and, once attached and recovered, they will no longer be susceptive to infection by the same strain. To achieve this, the body has a complicated network of diverse immune cells. Beretta and his colleagues summarize their study of modeling of an immune system dynamics in which time delay is incorporated.

In the last chapter, H.I. Freedman studies cancer, which originates from the self-cells of the patient, but which then become hostile by mutations. There is much in common between cancer cells and pathogens originated from outside of the host body. Freedman discusses the optimal chemotherapy, considering the cost and benefit of chemotherapy.

This collection of papers gives an overview of theoretical studies of infectious disease dynamics and evolution, and hopefully will serve as a source in future studies of different aspects of infectious disease dynamics. Here, the key words are time delay, spatial dynamics, and evolution.

Toward the end of this introductory chapter, we would like to note one limitation — all of the papers in this volume discuss deterministic models, which are accurate when the population size is very large. Since the number of microparasites, such as bacteria, or viruses, or cancer cells, is often very large, the neglect of stochasticity due to the finiteness of individuals seems to be acceptable. However, when we consider the speed of the appearance of novel mutants, we do need stochastic models, because mutants always start from a small number. According to studies on the timing of cancer initiation, which starts from rare mutations followed by population growth of cancer cells, the predictions of deterministic models differ by several orders of magnitude from those of stochastic models and direct computer simulations.

References

1. Anderson, R. M. and R. M. May (1991), *Infectious diseases of humans.* Oxford University Press, Oxford UK.
2. Kermack, W. O. and A. G. McKendrick (1927), A contribution to the mathematical theory of epidemics. *Proc. Roy. Soc. A* **115**, 700–721.
3. Shigesada, N. and K. Kawasaki (1997), Biological Invasions: Theory and Practice. Oxford University Press, Oxford.

Basic Knowledge and Developing Tendencies in Epidemic Dynamics

Zhien Ma and Jianquan Li

Summary. Infectious diseases have been a ferocious enemy since time immemorial. To prevent and control the spread of infectious diseases, epidemic dynamics has played an important role on investigating the transmission of infectious diseases, predicting the developing tendencies, estimating the key parameters from data published by health departments, understanding the transmission characteristics, and implementing the measures for prevention and control. In this chapter, some basic ideas of modelling the spread of infectious diseases, the main concepts of epidemic dynamics, and some developing tendencies in the study of epidemic dynamics are introduced, and some results with respect to the spread of SARS in China are given.

2.1 Introduction

Infectious diseases are those caused by pathogens (such as viruses, bacteria, epiphytes) or parasites (such as protozoans, worms), and which can spread in the population. It is well known that infectious diseases have been a ferocious enemy from time immemorial. The plague spread in Europe in 600 A.C., claiming the lives of about half the population of Europe (Brauer and Castillo-Chavez 2001). Although human beings have been struggling indomitably against various infections, and many brilliant achievements earmarked in the 20th century, the road to conquering infectious diseases is still tortuous and very long. Now, about half the population of the world (6 billion people) suffer the threat of various infectious diseases. For example, in 1995, a report of World Health Organization (WHO) shows that infectious diseases were still the number one of killers for human beings, claiming the lives of 52 million people in the world, of which 17 million died of various infections within that single year (WHO). In the last three decades, some new infectious diseases (such as Lyme diseases, toxic-shock syndrome, hepatitis C, hepatitis E) emerged. Notably, AIDS emerged in 1981 and became a deadly sexually transmitted disease throughout the world, and the newest Severe Acute Respiratory Syndrome (SARS) erupted in China in 2002, spreading to

31 countries in less than 6 months. Both history and reality show that, while human beings are facing menace from various infectious diseases, the importance of investigating the transmission mechanism, the spread rules, and the strategy of prevention and control is increasing rapidly, and such studies ar an important mission to be tackled urgently.

Epidemic dynamics is an important method of studying the spread rules of infectious diseases qualitatively and quantitatively. It is based largely on the specific properties of population growth, the spread rules of infectious diseases, and related social factors, serving to construct mathematical models reflecting the dynamical property of infectious diseases, to analyze the dynamical behavior qualitatively or quantitatively, and to carry out simulations. Such research results are helpful to predict the developing tendency of infectious diseases, to determine the key factors of spread of infectious diseases, and to seek the optimum strategy of preventing and controlling the spread of infectious diseases. In contrast with classic biometrics, dynamical methods can show the transmission rules of infectious diseases from the mechanism of transmission of the disease, so that we may learn about the global dynamical behavior of transmission processes. Incorporating statistical methods and computer simulations into epidemic dynamical models could make modelling methods and theoretical analyses more realistic and reliable, enabling us to understand the spread rules of infectious diseases more thoroughly.

The purpose of this article is to introduce the basic ideas of modelling the spread of infectious diseases, the main concepts of epidemic dynamics, some development tendencies of analyzing models of infectious diseases, and some SARS spreading models in China.

2.2 The fundamental forms and the basic concepts of epidemic models

2.2.1 The fundamental forms of the models of epidemic dynamics

Although Bernouilli studied the transmission of smallpox using a mathematical model in 1760 (Anderson and May 1982), research of deterministic models in epidemiology seems to have started only in the early 20th century. In 1906, Hamer constructed and analyzed a discrete model (Hamer 1906) to help understand the repeated occurrence of measles; in 1911, the Public Health Doctor Ross analyzed the dynamical behavior of the transmission of malaria between mosquitos and men by means of differential equation (Ross 1911); in 1927, Kermack and McKendrick constructed the famous compartmental model to analyze the transmitting features of the Great Plague which appeared in London from 1665 to 1666. They introduced a "threshold theory", which may determine whether the disease is epidemic or not (Kermack and McKendrick 1927, 1932), and laid a foundation for the research of epidemic dynamics. Epidemic dynamics flourished after the mid-20th century, Bailey's book being one of the landmark books published in 1957 and reprinted in 1975 (Baily 1975).

Kermack and McKendrick compartment models

In order to formulate the transmission of an epidemic, the population in a region is often divided into different compartments, and the model formulating the relations between these compartments is called **compartmental model**.

In the model proposed by Kermack and McKendrick in 1927, the population is divided into three compartments: a **susceptible compartment** labelled S, in which all individuals are susceptible to the disease; an **infected compartment** labelled I, in which all individuals are infected by the disease and have infectivity; and a **removed compartment** labelled R, in which all individuals are removed from the infected compartment. Let $S(t), I(t)$, and $R(t)$ denote the number of individuals in the compartments S, I, and R at time t, respectively. They made the following three assumptions:

1. The disease spreads in a closed environment (no emigration and immigration), and there is no birth and death in the population, so the total population remains constant, K, i. e., $S(t) + I(t) + R(t) = K$.
2. An infected individual is introduced into the susceptible compartment, and contacts sufficient susceptibles at time t, so the number of new infected individuals per unit time is $\beta S(t)$, where β is the **transmission coefficient**. The total number of newly infected is $\beta S(t)I(t)$ at time t.
3. The number removed (recovered) from the infected compartment per unit time is $\gamma I(t)$ at time t, where γ is the **rate constant for recovery**, corresponding to a mean infection period of $\frac{1}{\gamma}$. The recovered have permanent immunity.

For the assumptions given above, a compartmental diagram is given in Fig. 2.1. The compartmental model corresponding to Fig. 2.1 is the following:

$$\begin{cases} S' = -\beta SI \,, \\ I' = \beta SI - \gamma I \,, \\ R' = \gamma I \,. \end{cases} \tag{1}$$

Since there is no variable R in the first two equations of (1), we only need to consider the following equations

$$\begin{cases} S' = -\beta SI \,, \\ I' = \beta SI - \gamma I \end{cases} \tag{2}$$

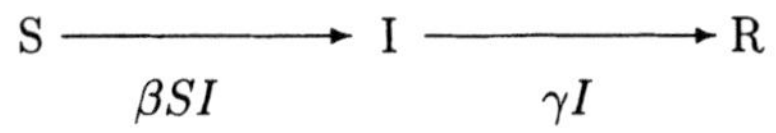

$$S \xrightarrow{\quad\beta SI\quad} I \xrightarrow{\quad\gamma I\quad} R$$

Fig. 2.1. Diagram of the SIR model without vital dynamics

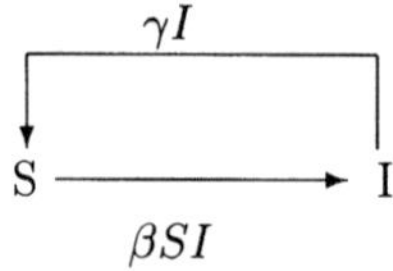

Fig. 2.2. Diagram of the SIS model without vital dynamics

to obtain the dynamic behavior of the susceptible and the infective. After that, the dynamic behavior of the removed R is easy to establish from the third equation of system (1), if necessary.

In general, if the disease comes from a virus (such as flu, measles, chicken pox), the recovered possess a permanent immunity. It is then suitable to use the SIR model (1). If the disease comes from a bacterium (such as cephalitis, gonorrhea), then the recovered individuals have no immunity, in other words, they can be infected again. This situation may be described using the SIS model, which was proposed by Kermack and McKendrick in 1932 (Kermack and McKendrick 1932). Its compartmental diagram is given in Fig. 2.2.

The model corresponding to Fig. 2.2 is

$$\begin{cases} S' = -\beta SI + \gamma I\,, \\ I' = \beta SI - \gamma I\,. \end{cases} \tag{3}$$

Up to this day, the idea of Kermack and McKendrick in establishing these compartmental models is still used extensively in epidemiological dynamics, and is being developed incessantly. According to the modelling idea, by means of the compartmental diagrams we list the fundamental forms of the model on epidemic dynamics as follows.

Models without vital dynamics

When a disease spreads through a population in a relatively short time, usually the births and deaths (vital dynamics) of the population may be neglected in the epidemic models, since the epidemic occurs relatively quickly, such as influenza, measles, rubella, and chickenpox.

(1) The models without the latent period

SI model In this model, the infected individuals can not recover from their illness, and the diagram is as follows:

S ⟶ I

βSI

SIS model In this model, the infected individuals can recover from the illness, but have no immunity.The diagram is shown in Fig. 2.2.

SIR model In this model, the removed individuals have permanent immunity after recovery. The diagram is shown in Fig. 2.1.

SIRS model In this model, the removed individuals have temporary immunity after recovery from the illness. Assume that due to the loss of immunity, the number of individuals being moved from the removed compartment to the susceptible compartment per unit time is $\delta R(t)$ at time t, where δ is the **rate constant for loss of immunity**, corresponding to a mean immunity period $\frac{1}{\delta}$. The diagram is as follows:

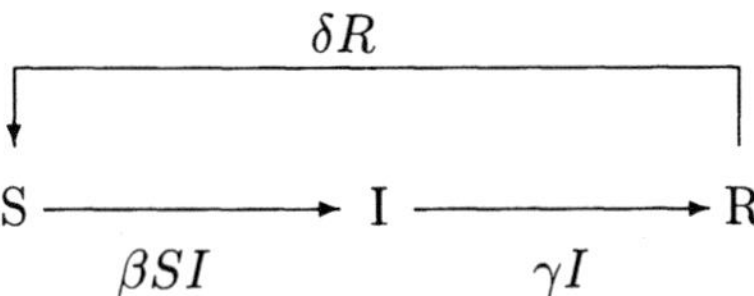

Remark 1. In the SIS model, the infected individuals may be infected again as soon as they recover from the infection. In the SIRS model, the removed individuals can not be infected in a given period of time, and may not be infected until they loose the immunity and become susceptible again.

(2) The models with the latent period

Here we introduce a new compartment, E (called **exposed compartment**), in which all individuals are infected but not yet infectious. The exposed compartment is often omitted, because it is not crucial for the susceptible-infective interaction or the latent period is relatively short.

Let $E(t)$ denote the number of individuals in the exposed compartment at time t. Corresponding to the model without the latent period, we can introduce some compartmental models such as SEI, SEIS, SEIR, and SEIRS. For example, the diagram of the SEIRS model is as follows:

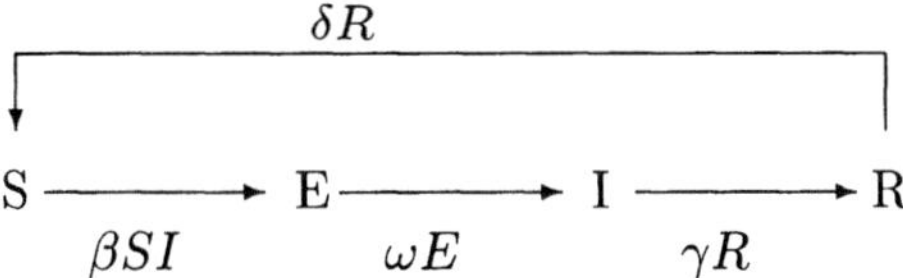

where ω is the **transfer rate constant** from the compartment E to the compartment I, corresponding to a mean latent period $\frac{1}{\omega}$.

Models with vital dynamics

(1) The size of the population is constant

If we assume that the birth and death rates of a population are equal while the disease actively spreads, and that the disease does not result in the death of the infected individuals, then the number of the total population is a constant, denoted by K. In the following, we give two examples for this case.

SIR model without vertical transmission In this model, we assume that the maternal antibodies can not be inherited by the infants, so all newborn infants are susceptible to the infection. Then, the corresponding compartmental diagram of the SIR model is as follows:

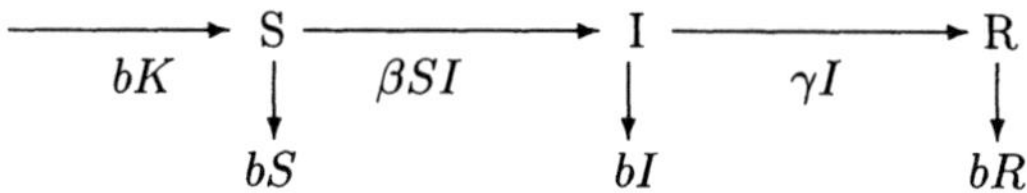

Fig. 2.3. Diagram of the SIR model without vertical transmission

SIR model with vertical transmission For many diseases, some newborn infants of infected parents are to be infected. This effect is called **vertical transmission**, such as AIDS, hepatitis B. We assume that the fraction k of infants born by infected parents is infective, and the rest of the infants are susceptible to the disease. Then, the corresponding compartmental diagram of the SIR model is as follows:

(2) The size of the population varies

When the birth and death rates of a population are not equal, or when there is an input and output for the total population, or there is death due to the infection, then the number of the total population varies. The number of the total population at time t is often denoted by $N(t)$.

SIS model with vertical transmission, input, output, and disease-related death The diagram is as follows:

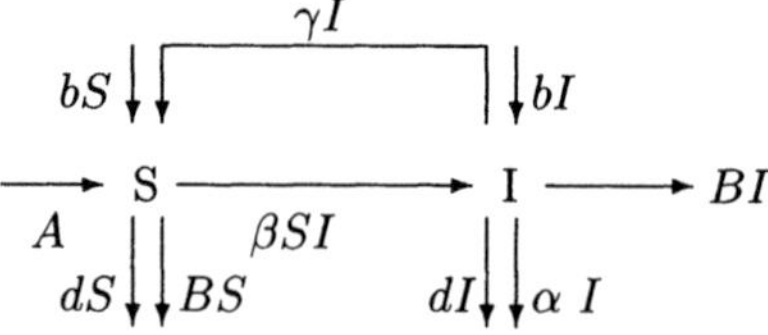

Fig. 2.4. Diagram of the SIS model with vertical transmission

Here, the parameter b represents the birth rate constant, d the natural death rate constant, α the death rate constant due to the disease, A the input rate of the total population, B the output rate constant of the susceptible and the infective.

MSEIR model with passive immunity Here, we introduce the compartment M in which all individuals are newborn infants with passive immunity. After the maternal antibodies disappear from the body, the infants move to the compartment S. Assume that the fraction of newborn infants with passive immunity is μ, and that the transfer rate constant from the compartment M to the compartment S is δ (corresponding to a mean period of passive immunity $\frac{1}{\delta}$). The diagram is as follows:

$$
\begin{array}{ccccccccc}
\mu bN & & (1-\mu)bN & & & & \alpha I & & \\
\downarrow & & \downarrow & & & & \uparrow & & \\
M & \longrightarrow & S & \longrightarrow & E & \longrightarrow & I & \longrightarrow & R \\
\downarrow \delta M & & \downarrow \beta SI & & \downarrow \omega E & & \downarrow \gamma R & & \downarrow \\
dM & & dS & & dE & & dI & & dR
\end{array}
$$

According to the diagrams shown above, we can easily write the corresponding compartmental models. For example, the SIR model corresponding to Fig. 2.3 is as follows:

$$
\begin{cases}
S' = bK - \beta SI - bS\,, \\
I' = \beta SI - bI - \gamma I\,, \\
R' = \gamma I - bR\,.
\end{cases}
\tag{4}
$$

The SIS model corresponding to Fig. 2.4 is as follows:

$$
\begin{cases}
S' = A + bS - \beta SI - dS - BS + \gamma I\,, \\
I' = bI + \beta SI - dI - \gamma I - BI - \alpha I\,.
\end{cases}
$$

2.2.2 The basic concepts of epidemiological dynamics

Adequate contact rate and incidence

It is well known that infections are transmitted through direct contact. The number of times an infective individual comes into contact with other members per unit time is defined as the **contact rate**, which often depends on the number N of individuals in the total population, and is denoted by a function $U(N)$. If the individuals contacted by an infected individual are susceptible, then they may be infected. Assuming that the probability of infection by every contact is β_0, then the function $\beta_0 U(N)$, called the **adequate contact rate**, shows the ability of an infected individual infecting others (depending on the environment, the toxicity of the virus or bacterium, etc.). Since, except for the susceptible, the individuals in other compartments of the population can not be infected when they make contact with the infectives, and the fraction of the susceptibles in the total population is $\frac{S}{N}$, then the mean adequate contact rate of an infective to the susceptible individuals is $\beta_0 U(N)\frac{S}{N}$, which is called the **infective rate**. Further, the number of new infected individuals resulting per unit time at time t is $\beta_0 U(N)\frac{S(t)}{N(t)}I(t)$, which is called the **incidence** of the disease.

When $U(N) = kN$, that is, the contact rate is proportional to the size of the total population, the incidence is $\beta_0 k S(t)I(t) = \beta S(t)I(t)$ (where $\beta = \beta_0 k$ is defined as the **transmission coefficient**), which is described as **bilinear incidence** or **simple mass-action incidence**. When $U(N) = k'$, that is, the contact rate is a constant, the incidence is $\beta_0 k'\frac{S(t)}{N(t)}I(t) = \frac{\beta S(t)I(t)}{N(t)}$ (where $\beta = \beta_0 k'$), which is described as **standard incidence**. For instance, the incidence formulating a sexually transmitted disease is often of standard type. The two types of incidence mentioned above are often used, but they are special for real cases. In recent years, some contact rates with saturation features between them were proposed, such as $U(N) = \frac{\alpha N}{1+\omega N}$ (Dietz 1982), $U(N) = \frac{\alpha N}{1+bN+\sqrt{1+2bN}}$ (Heesterbeek and Metz 1993). In general, the saturation contact rate $U(N)$ satisfies the following conditions:

$$U(0) = 0, \ U'(N) \geq 0, \ \left(\frac{U(N)}{N}\right)' \leq 0, \ \lim_{N\to\infty} U(N) = U_0 .$$

In addition, some incidences which are much more plausible for some special cases were also introduced, such as $\beta S^p I^q$, $\frac{\beta S^p I^q}{N}$ (Liu et al. 1986, 1987).

Basic reproduction number and modified reproduction number

In the following, we introduce two examples to understand the two concepts.

Example 1

We consider the SIS model (3) of Kermack and McKendrick. Since $S(t) + I(t) = K$(constant), (3) can be changed into the equation

$$S' = \beta(K - S)\left(\frac{\gamma}{\beta} - S\right) . \tag{5}$$

When $\frac{\gamma}{\beta} \geq K$, (5) has a unique equilibrium $S = K$ on the interval $(0, K]$ which is asymptotically stable, that is, the solution $S(t)$ starting from any $S_0 \in (0, K)$ increases to K as t tends to infinity. Meanwhile, the solution $I(t)$ decreases to zero. This implies that the infection dies out eventually and does not develop to an endemic.

When $\frac{\gamma}{\beta} < K$, (5) has two positive equilibria: $S = K$ and $S = \frac{\gamma}{\beta}$, where $S = K$ is unstable, and $S = \frac{\gamma}{\beta}$ is asymptotically stable. The solution $S(t)$ starting from any $S_0 \in (0, K)$ approaches to $\frac{\gamma}{\beta}$ as t tends to infinity, and $I(t)$ tends to $K - \frac{\gamma}{\beta} > 0$. Thus, point $\left(\frac{\gamma}{\beta}, K - \frac{\gamma}{\beta}\right)$ in S-I plane is called the **endemic equilibrium** of system (3). This case is not expected.

Therefore, $\frac{\gamma}{\beta} = K$, i.e., $R_0 := \frac{\beta K}{\gamma} = 1$ is a threshold which determines whether the disease dies out ultimately. The disease dies out if $R_0 < 1$, is endemic if $R_0 > 1$.

The epidemiological meaning of R_0 as a threshold is intuitively clear. Since $\frac{1}{\gamma}$ is the mean infective period, and βK is the number of new cases infected per unit time by an average infective which is introduced into the susceptible compartment in the case that all the members of the population are susceptible, i.e., the number of individuals in the susceptible compartment is K (this population is called a completely susceptible population), then R_0 represents the average number of secondary infections that occur when an infective is introduced into a completely susceptible population. So, $R_0 < 1$ implies that the number of infectives tends to zero, and $R_0 > 1$ implies that the number of infectives increases. Hence, the threshold R_0 is called the **basic reproduction number**.

Example 2

Consider an SIR model with exponential births and deaths and the standard incidence. The compartmental diagram is as follows:

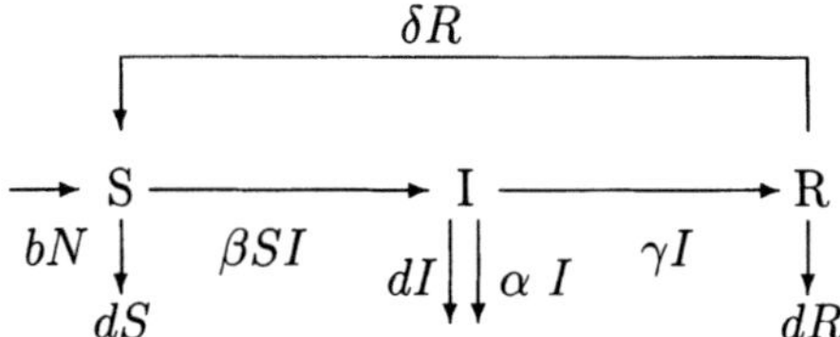

The differential equations for the diagram are

$$\begin{cases} S' = bN - dS - \frac{\beta SI}{N} + \delta R \,, \\ I' = \frac{\beta SI}{N} - (\alpha + d + \gamma)I \,, \\ R' = \gamma I - (d + \delta)R \,, \end{cases} \tag{6}$$

where b is the birth rate constant, d the natural death rate constant, and α the disease-related death rate constant.

Let $N(t) = S(t) + I(t) + R(t)$, which is the number of individuals of total population, and then from (6), $N(t)$ satisfies the following equation:

$$N' = (b - d)N - \alpha I \,. \tag{7}$$

The net growth rate constant in a disease-free population is $r = b - d$. In the absence of disease (that is, $\alpha = 0$), the population size $N(t)$ declines exponentially to zero if $r < 0$, remains constant if $r = 0$, and grows exponentially if $r > 0$. If disease is present, the population still declines to zero if $r < 0$. For $r > 0$, the population can go to zero, remain finite or grow exponentially, and the disease can die out or persist.

On the other hand, we may determine whether the disease dies out or not by analyzing the change tendency of the infective fraction $\frac{I(t)}{N(t)}$ in the total population. If $\lim\limits_{t\to\infty} \frac{I(t)}{N(t)}$ is not equal to zero, then the disease persists; if $\lim\limits_{t\to\infty} \frac{I(t)}{N(t)}$ is equal to zero, then the disease dies out.

Let

$$x = \frac{S}{N}, \quad y = \frac{I}{N}, \quad z = \frac{R}{N},$$

then x, y, and z represent the fractions of the susceptible, the infective, and the removed in the total population. From (6) and (7) we have

$$\begin{cases} x' = b - bx - \beta xy + \delta z + \alpha xy, \\ y' = \beta xy - (b + \alpha + \gamma)y + \alpha y^2, \\ z' = \gamma y - (b + \delta)z + \alpha yz, \end{cases} \tag{8}$$

which is actually a two-dimensional system due to $x + y + z = 1$.

Substituting $x = 1 - y - z$ into the middle equation of (8) gives the equations

$$\begin{cases} y' = \beta(1 - y - z)y - (b + \alpha + \gamma)y + \alpha y^2, \\ z' = \gamma y - (b + \delta)z + \alpha yz. \end{cases} \tag{9}$$

Let

$$R_1 = \frac{\beta}{b + \alpha + \gamma}.$$

It is easy to verify that when $R_1 \leq 1$, (9) has only the equilibrium $P_0(0,0)$ (called **disease-free equilibrium**) in the feasible region which is globally asymptotically stable; when $R_1 > 1$, (9) has the disease-free equilibrium $P_0(0,0)$ and the positive equilibrium $P^*(y^*, z^*)$ (called the endemic equilibrium), where P_0 is unstable and P^* is globally asymptotically stable (Busenberg and Van den driessche 1990).

The fact that the disease-free equilibrium P_0 is globally asymptotically stable implies $\lim\limits_{t\to\infty} y(t) = \lim\limits_{t\to\infty} \frac{I(t)}{N(t)} = 0$, i. e., the infective fraction goes to zero. In this sense, the disease dies out finally no matter what the total population size $N(t)$ keeps finite, goes to zero or grows infinitely. The fact that the endemic equilibrium P^* is globally asymptotically stable implies $\lim\limits_{t\to\infty} y(t) = \lim\limits_{t\to\infty} \frac{I(t)}{N(t)} = y^* > 0$, i. e., the infective fraction goes to a positive constant. This shows that the disease persists in population.

It is seen from (6) that the mean infectious period is $\frac{1}{d+\alpha+\gamma}$, the incidence is of standard type, and the adequate contact rate is β, so that the basic reproduction number of model (6) is $R_0 = \frac{\beta}{d+\alpha+\gamma}$.

From the results above we can see that, for this case, the threshold to determine whether the disease dies out is $R_1 = 1$ but not $R_0 = 1$. Therefore, the number R_1 is defined as **modified reproduction number**.

2.3 Some tendencies in the development of epidemic dynamics

2.3.1 Epidemic models described by ordinary differential equations

So far, many results in studying epidemic dynamics have been achieved. Most models involve ordinary equations, such as the models listed in Sect. 2.2.1. When the total population size is a constant, the models SIS, SIR, SIRS and SEIS can be easily reduced to a plane differential system, and the results obtained are often complete. When the birth and death rates of the population are not equal, or the disease is fatal, etc., the total population is not a constant, so that the model can not be reduced in dimensions directly, and the related investigation becomes complex and difficult. Though many results have been obtained by studying epidemic models with bilinear and standard incidences, most of these are confined to local dynamic behavior, global stability is often obtained only for the disease-free equilibrium, and the complete results with respect to the endemic equilibrium are limited.

In the following, we introduce some epidemic models described by ordinary differential equations and present some common analysis methods, and present some related results.

SIRS model with constant input and exponent death rate and bilinear incidence

We first consider the model

$$\begin{cases} S' - A - dS - \beta SI + \delta R \,, \\ I' = \beta SI - (\alpha + d + \gamma)I \,, \\ R' = \gamma I - (d + \delta)R \,. \end{cases} \tag{10}$$

Let $N(t) = S(t) + I(t) + R(t)$, then from (10) we have

$$N'(t) = A - dN - \alpha I \,, \tag{11}$$

and thus it is easy to see that the set

$$D = \left\{ (S, I, R) \in R^3 \middle| 0 < S + I + R \leq \frac{A}{d}, S > 0, I \geq 0, R \geq 0 \right\}$$

is a positive invariant set of (10).

Theorem 1. *(Mena-Lorca and Hethcote 1992) Let* $R_0 = \frac{A\beta}{d(\gamma+\alpha+d)}$. *The disease-free equilibrium* $E_0\left(\frac{A}{d}, 0, 0\right)$ *is globally asymptotically stable on the set D if $R_0 \leq 1$ and unstable if $R_0 > 1$. The endemic equilibrium $E^*(x^*, y^*, z^*)$ is locally asymptotically stable if $R_0 > 1$. Besides, when $R_0 > 1$, the endemic equilibrium E^* is globally asymptotically stable for the case $\delta = 0$.*

Proof

The global stability of the disease-free equilibrium E_0 can be easily proved by using the Liapunov function $V = I$, LaSalle's invariance principle, and the theory of limit system if $R_0 \leq 1$.

In order to prove the stability of the endemic equilibrium E^* for $R_0 > 1$, we make the following variable changes:

$$S = S^*(1 + x), \quad I = I^*(1 + y), \quad R = R^*(1 + z),$$

then (10) becomes

$$\begin{cases} x' = -\beta I^* \left[\left(\frac{\delta}{\beta I^*} + 1 \right) x + y + xy \right], \\ y' = \beta S^* x(1 + y), \\ z' = d(y - z). \end{cases} \tag{12}$$

Noticing that the origin of (12) corresponds to the endemic equilibrium E^* of (10), we define the Liapunov function

$$V = \frac{x^2}{2\beta I^*} + \frac{1}{\beta S^*} [y - \ln(1 + y)],$$

and then the derivative of V along the solution of (12) is

$$V'|_{(12)} = -x^2 \left(\frac{\delta}{\beta I^*} + 1 + y \right) \leq 0.$$

Thus, the global asymptotical stability of the origin of (12) (i. e., the endemic equilibrium E^*) can be obtained by using LaSalle's invariance principle.

Models with latent period

In general, some models with latent period (such as SEIR and SEIRS) may not be reduced to plane differential systems, but they may be competitive systems under some conditions. In this case, the global stability of some of these models can be obtained by means of the orbital stability, the second additive compound matrix, and the method of ruling out the existence of periodic solutions proposed by Muldowney and Li (Li and Muldowney 1995, 1996; Muldowney 1990).

For example, the SEIR model with constant input and bilinear incidence

$$\begin{cases} S' = A - dS - \beta SI, \\ E' = \beta SI - (\epsilon + d)E, \\ I' = \epsilon E - (\gamma + \alpha + d)I, \\ R' = \gamma I - dR \end{cases} \tag{13}$$

has the following results:

Theorem 2. *(Li and Wang to appear) Let $R_0 = \frac{A\beta\epsilon}{d(d+\epsilon)(d+\gamma+\alpha)}$. The disease-free equilibrium is globally asymptotically stable if $R_0 \leq 1$ and unstable if $R_0 > 1$; the unique endemic equilibrium is globally asymptotically stable if $R_0 > 1$.*

For the SEIR model with exponent input and standard incidence

$$\begin{cases} S' = bN - dS - \frac{\beta SI}{N}\,, \\ E' = \frac{\beta SI}{N} - (\epsilon + d)E\,, \\ I' = \epsilon E - (\gamma + \alpha + d)I\,, \\ R' = \gamma I - dR\,, \end{cases}$$

Li et al. (1999) introduced the fraction variables: $s = \frac{S}{N}, e = \frac{E}{N}, i = \frac{I}{N}$ and $r = \frac{R}{N}$, and they obtained the following results:

Theorem 3. *(Li and et al. 1999) Let $R_0 = \frac{\beta\epsilon}{(d+\epsilon)(d+\gamma+\alpha)}$. The disease-free equilibrium is globally asymptotically stable if $R_0 \leq 1$ and unstable if $R_0 > 1$; the unique endemic equilibrium is locally asymptotically stable if $R_0 > 1$ and globally asymptotically stable if $R_0 > 1$ and $\alpha \leq \epsilon$.*

Recently, Zhang and Ma (2003) applied the saturation incidence

$$C(N) = \frac{bN}{1 + bN + \sqrt{1 + 2bN}}$$

instead of the bilinear one in (13), analyzed its global stability completely by using analogous methods, and obtained the basic reproduction number of the corresponding model

$$R_0 = \frac{\beta b\epsilon A}{(d + \gamma + \alpha)(d + \epsilon)(d + bA + \sqrt{d^2 + 2bdA})}\,.$$

For the SEIS model with constant input, Fan et al. (2001) obtained the complete global behavior with respect to the bilinear incidence. Zhang and Ma (to appear) generalized the incidence to the general form $\beta C(N)\frac{SI}{N}$, where $C(N)$ satisfies the following conditions: (1) $C(N)$ is non-negative, non-decreasing, and continuous differentiable with respect to N; (2) $\frac{C(N)}{N}$ is non-increasing and continuous differentiable with respect to $N > 0$, and obtained the following results by similar methods for the model

$$\begin{cases} S' = dK - dS - \beta C(N)\frac{SI}{N}\,, \\ E' = \beta C(N)\frac{SI}{N} - (\epsilon + d)E\,, \\ I' = \epsilon E - (\gamma + \alpha + d)I\,, \\ R' = \gamma I - dR\,, \end{cases}$$

Theorem 4. *(Zhang and Ma to appear) Let $R_0 = \frac{\beta\epsilon C(K)}{(d+\epsilon)(d+\gamma+\alpha)}$. The disease-free equilibrium is globally asymptotically stable if $R_0 \leq 1$ and unstable if $R_0 > 1$; the unique endemic equilibrium is globally asymptotically stable if $R_0 > 1$.*

The similar method is also used to discuss the global behavior of an SEIR model with bilinear incidence and vertical transmission (Li et al. 2001).

Models with quarantine of the infectives

So far, there are two effective measures to control and prevent the spread of the infection, these being quarantine and vaccination. The earliest studies on the effects of quarantine on the transmission of the infection was achieved by Feng and Thieme (2003a, 2003b), and Wu and Feng (2000). In those papers, they introduced a quarantined compartment, Q, and assumed that all infective individuals must pass through the quarantined compartment before going to the removed compartment or back to the susceptible compartment. Hethcote, Ma, and Liao (2002) considered more realistic cases: a part of the infectives are quarantined, whereas the others are not quarantined and enter into the susceptible compartment or into the removed compartment directly. They analyzed six SIQS and SIQR models with bilinear, standard or quarantine-adjusted incidence, and found that for the SIQR model with quarantine-adjusted incidence, the periodic solutions may arise by Hopf bifurcation, but for the other five models with disease-related death, sufficient and necessary conditions assuring the global stability of the disease-free equilibrium and the endemic equilibrium were obtained.

For instance, for the SIQS model with bilinear incidence

$$\begin{cases} S' = A - \beta SI - dS + \gamma I + \epsilon Q \,, \\ I' = [\beta S - (d + \alpha + \delta + \gamma)]I \,, \\ Q' = \delta I - (d + \alpha + \epsilon)Q \,, \end{cases} \tag{14}$$

the following hold:

Theorem 5. *(Hethcote, Ma, and Liao 2002) Let $R_q = \frac{A\beta}{d(\gamma + \delta + d + \alpha)}$. The disease-free equilibrium is globally asymptotically stable if $R_q \leq 1$ and unstable if $R_q > 1$; the unique endemic equilibrium is globally asymptotically stable if $R_q > 1$.*

To prove the global stability of the endemic equilibrium of system (14), let $N(t) = S(t) + I(t) + Q(t)$, and then system (14) becomes the system

$$\begin{cases} N' = -d(N - N^*) - \alpha\,(I - I^*) - \alpha\,(Q - Q^*) \,, \\ I' = \beta[(N - N^*) - (I - I^*) - (Q - Q^*)]I \,, \\ Q' = \delta(I - I^*) - (d + \alpha + \epsilon)(Q - Q^*) \,, \end{cases}$$

where the point (N^*, I^*, Q^*) is the endemic equilibrium of this last system. Define the Liapunov function

$$\begin{aligned} V(N, I, Q) ={}& \frac{\delta + \epsilon + 2d + \alpha}{\beta} \left[I - I^* - I^* \ln \frac{I}{I^*} \right] \\ &+ \frac{1}{2} \left\{ \frac{(\epsilon + 2d)(N - N^*)^2}{\alpha} + \left[(N - N^*) - (Q - Q^*) \right]^2 \right. \\ &\left. + \frac{(\epsilon + 2d)(Q - Q^*)^2}{\delta} \right\} \,, \end{aligned}$$

then the global stability of the endemic equilibrium can be obtained by computing the derivative of $V(N, I, Q)$ along the solution of the system.

Since the quarantined individuals do not come into contact with the unquarantined individuals, for the case that the adequate contact rate is constant, the incidence should be $\frac{\beta SI}{N-Q} = \frac{\beta SI}{S+I+R}$, which is called the quarantine-adjusted incidence. Then, the SIQR model with quarantine-adjusted incidence is

$$\begin{cases} S' = A - \frac{\beta SI}{S+I+R} - dS\,, \\ I' = \frac{\beta SI}{S+I+R} - (\gamma + \delta + d + \alpha)I\,, \\ Q' = \delta I - (d + \alpha_1 + \epsilon)Q\,, \\ R' = \gamma I + \epsilon Q - dR\,. \end{cases} \tag{15}$$

For (15), we have

Theorem 6. *(Hethcote, Ma, and Liao 2002) Let $R_q = \frac{\beta}{\gamma+\delta+d+\alpha}$. The disease-free equilibrium is globally asymptotically stable if $R_q \leq 1$ and unstable if $R_q > 1$. If $R_q > 1$, the disease is uniformly persistent, and (15) has a unique endemic equilibrium P^* which is usually locally asymptotically stable, but Hopf bifurcation can occur for some parameters, so that P^* is sometimes an unstable spiral and periodic solutions around P^* can occur.*

From Theorem 5 and Theorem 6, we know that the basic reproduction numbers of (14) and (15) include the recovery rate constant γ and the quarantined rate constant δ besides the disease-related death rate constant α, but do not include the recovery rate constant ϵ and the disease-related death rate constant α_1 of the quarantined. This implies that quarantining the infectives and treating the un-quarantined are of the same significance for controlling and preventing the spread of the disease, but this is not related to the behavior of the quarantined.

Models with vaccination Vaccinating the susceptible against the infection is another effective measure to control and prevent the transmission of the infection. To model the transmission of the infection under vaccination, ordinary differential equations, delay differential equations and pulse differential equation (Li and Ma 2002, 2003, 2004a, 2004b, to appear; Li et al. to appear; Jin 2001) are often used. Here, we only introduce some results of ordinary differential equations obtained by Li and Ma (2002, to appear).

The transfer diagram of the SISV model with exponential input and vaccination is

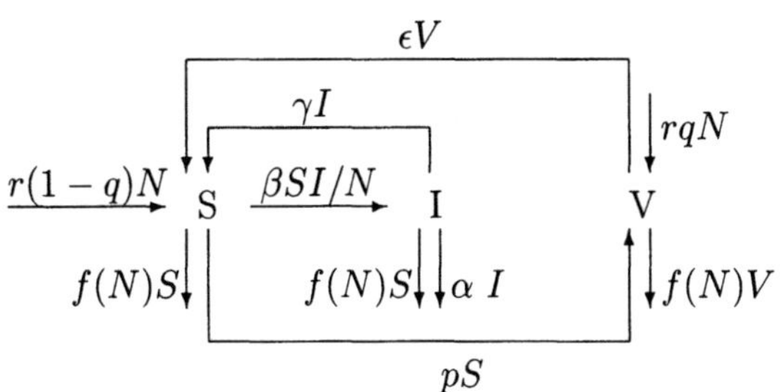

The model corresponding to the diagram is

$$\begin{cases} S' = r(1-q)N - \beta\frac{SI}{N} - [p + f(N)]S + \gamma I + \epsilon V , \\ I' = \beta\frac{SI}{N} - (\gamma + \alpha + f(N))I , \\ V' = rqN + pS - [\epsilon + f(N)]V , \end{cases} \qquad (16)$$

where V represents the vaccinated compartment. We assume that the vaccinated individuals have temporary immunity, the mean period of immunity is $\frac{1}{\epsilon}$, and that the natural death rate $f(N)$ depends on the total population N, which satisfies the following conditions:

$$f(N) > 0 , \quad f'(N) > 0 \quad \text{for } N > 0 \text{ and } f(0) = 0 < r < f(\infty) ,$$

where q represents the fraction of the vaccinated newborns, and p is the fraction of the vaccinated susceptibles. The other parameters have the same definitions as in the previous sections.

For system (16), by initially making the normalizing transformation to S, I and V , and then using the extensive Bendixson-Dulac Theorem of Ma et al. (2004), we can obtain the following results.

Theorem 7. *(Li and Ma 2002) Let $R_V = \frac{\beta[q+r(1-q)]}{(p+\epsilon+r)(\alpha+r+\gamma)}$. The disease-free equilibrium is globally asymptotically stable if $R_V \leq 1$ and unstable if $R_V > 1$; the unique endemic equilibrium is globally asymptotically stable if $R_V > 1$.*

For the model without vaccination (i. e., $p = q = 0$), the basic reproduction number of (16) is $R_0 = \frac{\beta}{\alpha+r+\gamma}$. By comparing R_V and R_0, Li and Ma (2002) came to the following conclusion: To control and prevent the spread of the disease, increasing the fraction q of the vaccinated newborns is more efficient when $rR_0 > 1$; increasing the fraction p of the vaccinated susceptibles is more efficient when $rR_0 < 1$.

Model (16) assumed that the vaccine is completely efficient, but, in fact, the efficiency of a vaccine is usually not 100%. Hence, incorporating the efficiency of the vaccine into epidemic models with vaccination is necessary. If we consider an SIS model with the efficiency of vaccine, then the system (16) will be changed into the following:

$$\begin{cases} S' = r(1-q)N - \beta\frac{SI}{N} - [p + f(N)]S + \gamma I + \epsilon V , \\ I' = \beta(S + \sigma V)\frac{I}{N} - (\gamma + \alpha + f(N))I , \\ V' = rqN + pS - \sigma\beta\frac{IV}{N} - [\epsilon + f(N)]V , \end{cases} \qquad (17)$$

where the fraction $\sigma(0 \leq \sigma \leq 1)$ reflects the inefficiency of the vaccination. The more effective the vaccine is, the less the value of σ is. Moreover, $\sigma = 0$ implies that the vaccine is completely effective in preventing infection, while $\sigma = 1$ implies that the vaccine is absolutely of no effect.

For model (17), we found the modified reproduction number

$$R_V = \frac{\beta[\epsilon + \sigma p + r(1 - (1 - \sigma)q)]}{(\alpha + r + \gamma)(p + \epsilon + r)} .$$

Theorem 8. *(Li, Ma and Zhou to appear)   For system (17), the following results are true.*

(1) When $R_V > 1$, there exists a unique endemic equilibrium which is globally asymptotically stable.

(2) When $R_V = 1, \alpha < \sigma\beta, B > 0$, there exists a unique endemic equilibrium which is globally asymptotically stable.

(3) When $R_V < 1, \alpha < \sigma\beta, \beta > r + \alpha + \gamma, B > 2\sqrt{AC}$, there exist two endemic equilibria: one is an asymptotically stable node, another is a saddle point.

(4) When $R_V < 1, \alpha < \sigma\beta, \beta > r + \alpha + \gamma, B = 2\sqrt{AC}$, there exists a unique endemic equilibrium, which is a saddle-node.

(5) For all other cases of parameters, the disease-free equilibrium is globally asymptotically stable.

Where

$$A = (\alpha - \sigma\beta)(\beta - \alpha), \quad C = (p + r + \epsilon)(r + \alpha + \gamma)(R_0 - 1),$$
$$B = \alpha(p + \epsilon + \gamma + \alpha + 2r) - \beta[(\alpha + r + \epsilon) - \sigma(\beta - r - \alpha - \gamma - p)].$$

According to Theorem 8, the change of endemic equilibrium of the system (17) can be shown in Fig. 2.5, while the common case is shown in Fig. 2.6. Figure 2.5 shows that, when R_V is less than but close to 1, the system (17) has two endemic equilibria, and has no endemic equilibrium until $R_V < R_c < 1$. One of these two equilibria is an asymptotically stable node, and the other is a saddle point. It implies that, for this case, whether the disease does die out or not depends on the initial condition. This phenomenon is called **backward bifurcation**. Therefore, incorporating the efficiency of vaccine into the epidemic models is important and necessary.

Within this context, the bifurcations with respect to epidemic models were also investigated by many researchers. Liu et al. (1986, 1987) discussed codimension one bifurcation in SEIRS and SIRS models with incidence $\beta I^p S^q$. Lizana and Rivero (1996) considered codimension two bifurcation in the SIRS model. Wu and Feng (2000) analyzed the homoclinic bifurcation in the SIQR model. Watmaough and van den Driessche (2000), Hadeler and van den

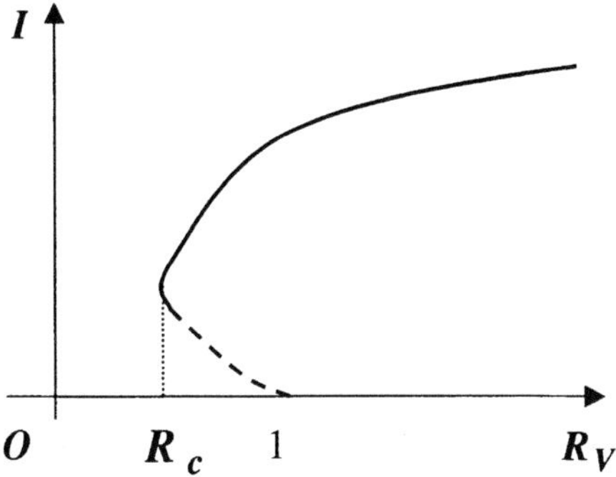

Fig. 2.5. Backward bifurcation

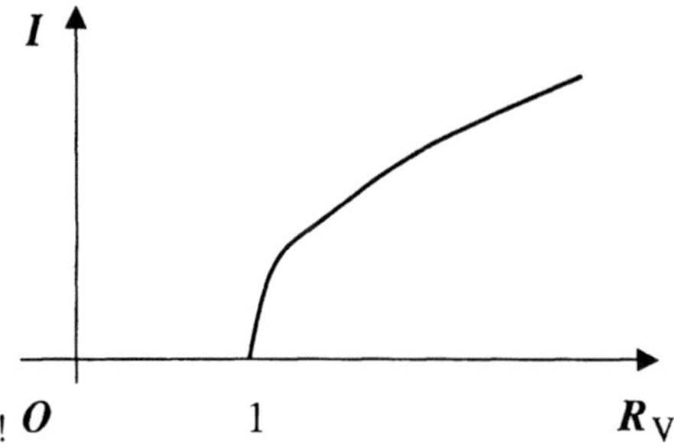

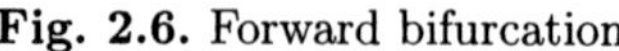
Fig. 2.6. Forward bifurcation

Driessche (1997), and Dushoff et al. (1998) discussed the backward bifurcation in some epidemic models. Ruan and Wang (2003) found the Bogdanov-Takens bifurcation in the SIRS model with incidence $\frac{kI^l S}{1+\alpha\ I^h}$.

2.3.2 Epidemic models with time delay

The models with time delay reflect the fact: the dynamic behavior of transmission of the disease at time t depends not only on the state at time t but also on the state in some period before time t.

Idea of modelling To formulate the idea of modelling the spread of disease, we show two SIS models with fixed delay (also called discrete delay) and distributed delay (also called continuous delay), respectively.

(1) Models with discrete delay

Assuming that the infective period of all the infectives is constant τ, then the rate at which the infectives recover and return to the susceptible compartment is $\beta S(t-\tau)I(t-\tau)$ if the rate of new infections at time t is $\beta S(t)I(t)$. Corresponding to the system (3), we have the SIS model with delay as follows

$$\begin{cases} S'(t) = -\beta S(t)I(t) + \beta S(t-\tau)I(t-\tau)\,, \\ I'(t) = \beta S(t)I(t) - \beta S(t-\tau)I(t-\tau)\,. \end{cases}$$

If the natural death rate constant d and the disease-related death rate constant α of the infectives are incorporated in the model, then the rate of recovery at time t should be $\beta S(t-\tau)I(t-\tau)e^{-(d+\alpha)\tau}$, where the factor $e^{-(d+\alpha)\tau}$ denotes the fraction of those individuals who were infected at time $t-\tau$ and survive until time t. Thus, we have the model

$$\begin{cases} S'(t) = -\beta S(t)I(t) - dS(t) + \beta S(t-\tau)I(t-\tau)e^{-(d+\alpha)\tau}\,, \\ I'(t) = \beta S(t)I(t) - (d+\alpha)I(t) - \beta S(t-\tau)I(t-\tau)e^{-(d+\alpha)\tau}\,. \end{cases}$$

(2) Models with distributed delay

The case that all the infectives have the same period of infection is an extreme one. In fact, the infective period usually depends on the difference of infected individuals. Assume that $P(\tau)$ is the probability that the individuals, who were infected at time 0, remain in the infected compartment at time τ. It is obvious that $P(0) = 1$. Thus, the number of the infectives at time t is

$$I(t) = \int_0^{+\infty} \beta S(t-\tau)I(t-\tau)P(\tau)\,\mathrm{d}\tau = \int_{-\infty}^{t} \beta S(u)I(u)P(t-u)\,\mathrm{d}u\,.$$

Assuming that $P(\tau)$ is differentiable, then from the last equation we have

$$I'(t) = \beta S(t)I(t) + \int_{-\infty}^{t} \beta S(u)I(u)P'(t-u)\,\mathrm{d}u\,.$$

Let $f(\tau) := -P'(\tau)$, then

$$I'(t) = \beta S(t)I(t) - \int_0^{\infty} \beta S(t-\tau)I(t-\tau)f(\tau)\,\mathrm{d}t\,.$$

It is easy to see that $\int_0^{+\infty} f(\tau)\,\mathrm{d}\tau = \int_0^{+\infty}[-P'(\tau)]\,\mathrm{d}\tau = 1$, and that $\int_0^{+\infty} \tau P(\tau)\,\mathrm{d}\tau$ is the infective period. Therefore, the corresponding SIS model is

$$\begin{cases} S'(t) = -\beta S(t)I(t) + \int_0^{+\infty} \beta S(t-\tau)I(t-\tau)f(\tau)\,\mathrm{d}\tau\,, \\ I'(t) = \beta S(t)I(t) - \int_0^{+\infty} \beta S(t-\tau)I(t-\tau)f(\tau)\,\mathrm{d}\tau\,. \end{cases}$$

Similarly to the case with the discrete delay, if the natural death rate constant d and the disease-related death rate constant α of the infectives are incorporated in the model, then the corresponding SIS model becomes

$$\begin{cases} S'(t) = -\beta S(t)I(t) - dS(t) + \int_0^{+\infty} \beta S(t-\tau)I(t-\tau)f(\tau)e^{-(d+\alpha)\tau}\,\mathrm{d}\tau\,, \\ I'(t) = \beta S(t)I(t) - (d+\alpha)I(t) - \int_0^{+\infty} \beta S(t-\tau)I(t-\tau)f(\tau)e^{-(d+\alpha)\tau}\,\mathrm{d}\tau\,. \end{cases}$$

In the following, we give some models established according to the idea above.

Example 3

Supposing that the birth and natural death of the population are of exponential type, the disease-related death rate constant is α, the infective period is a constant τ, and that there is no vertical transmission, then the SIS model with standard incidence is

$$\begin{cases} S'(t) = bN(t) - dS(t) - \dfrac{\beta S(t)I(t)}{N(t)} + \dfrac{\beta S(t-\tau)I(t-\tau)}{N(t-\tau)}e^{-(d+\alpha)\tau}\,, \\ I'(t) = \dfrac{\beta S(t)I(t)}{N(t)} - (d+\alpha)I(t) - \dfrac{\beta S(t-\tau)I(t-\tau)}{N(t-\tau)}e^{-(d+\alpha)\tau}\,, \end{cases}$$

where $N(t) = S(t) + I(t)$ satisfies the equation

$$N'(t) = (b - \mathrm{d})N(t) - \alpha\,I(t)\,.$$

Example 4

Let A be the birth rate of the population, d the natural death rate constant, α the disease-related death rate constant, ω the latent period, τ the infective period, then the SEIR model with bilinear incidence is

$$
\begin{cases}
S'(t) = A - dS(t) - \beta S(t)I(t) \,, \\
E'(t) = \beta S(t)I(t) - \beta S(t - \omega)I(t - \omega)\mathrm{e}^{-d\omega} - dE(t) \,, \\
I'(t) = \beta S(t - \omega)I(t - \omega)\mathrm{e}^{-d\omega} \\
\qquad\quad - \beta S(t - \omega - \tau)I(t - \omega - \tau)\mathrm{e}^{-d(\omega+\tau)}\mathrm{e}^{-\alpha\tau} - (d + \alpha)I(t) \,, \\
R'(t) = \beta S(t - \omega - \tau)I(t - \omega - \tau)\mathrm{e}^{-d(\omega+\tau)}\mathrm{e}^{-\alpha\tau} - dR(t) \,.
\end{cases}
$$

Example 5

If we incorporate the vaccination into the SIR model, and assume that the efficient rate constant of the vaccination is p, and that the efficient period of vaccine in the vaccinated body is a constant τ, then the SIR model with vaccination and bilinear incidence is the following:

$$
\begin{cases}
S'(t) = A - \beta I(t)S(t) - (d + p)S(t) + \gamma I(t) + pS(t - \tau)\mathrm{e}^{-d\tau} \,, \\
I'(t) = \beta I(t)S(t) - (d + \gamma + \alpha)I(t) \,, \\
R'(t) = \gamma I(t) + pS(t) - pS(t - \tau)\mathrm{e}^{-d\tau} - dR(t) \,.
\end{cases}
\tag{18}
$$

Note that the term $pS(t-\tau)\mathrm{e}^{-d\tau}$ in (18) represents the number of individuals who are vaccinated at time $t-\tau$ and still survive at time t, and the occurrence of delay form is due to the fact that the efficient period of vaccine is a fixed constant τ. If the probability of losing immunity is an exponential distribution $\mathrm{e}^{-\mu t}$ ($\frac{1}{\mu}$ is the mean efficient period of vaccine), then the corresponding model is a system of ordinary differential equations (Ma et al. 2004).

For delay differential systems, the local stability of equilibrium is often obtained by discussing the corresponding characteristic equation, which is similar to an ordinary differential equation. Also, the method to obtain the global stability is mainly to construct Liapunov functionals. For example, since the first two equations in (18) do not include the variable R obviously, we can only consider the subsystem consisting of the first two equations to obtain the following results:

Theorem 9. *(Li and Ma to appear) Let*

$$
R_V = \frac{\beta A}{(d + \alpha + \gamma)[d + p(1 - \mathrm{e}^{-d\tau})]} = \frac{\beta S_0}{d + \alpha + \gamma} \,.
$$

The disease-free equilibrium is globally asymptotically stable on the positively invariant set $D = \{(S, I) : S > 0, I \geq 0, S + I \leq \frac{A}{d}\}$ if $R_V \leq 1$ and unstable if $R_V > 1$. The unique endemic equilibrium is globally asymptotically stable in the positively invariant set D if $R_V > 1$.

The global stability of the disease-free equilibrium can be proved by constructing the Liapunov functional

$$V = \frac{(S - S_0)^2}{\alpha} + S_0 I + \frac{p\,\mathrm{e}^{-d\tau}}{2} \int_{t-\tau}^{t} (S(u) - S_0)^2 \, \mathrm{d}u \; ;$$

and the global stability of endemic equilibrium can be proved by constructing the Liapunov functional

$$V = \frac{(S - S_0)^2}{2} + \frac{d + \alpha + \gamma}{\beta} \left(I - I^* - I^* \ln \frac{I}{I^*} \right)$$
$$+ \frac{p\,\mathrm{e}^{-d\tau}}{2} \int_{t-\tau}^{t} (S(u) - S^*)^2 \, \mathrm{d}u \, ,$$

where S^* and I^* are the coordinates of the endemic equilibrium of the system (18).

For epidemic dynamical models with delay, many results have been obtained (Hethcote and van den Driessche 1995; Ma et al. 2002; Wang 2002; Wang and Ma 2002; Xiao and Chen 2001a; Yuan and Ma 2001, 2002; Yuan et al. 2003a, 2003b) , but few results were achieved with respect to the global stability of the endemic equilibrium. Especially the results about necessary and sufficient conditions like Theorem 9 are rare.

2.3.3 Epidemic models with age structure

Age has been recognized as an important factor in the dynamics of population growth and epidemic transmission, because individuals have usually different dynamic factors (such as birth and death) in different periods of age, and age structure also affects the transmission of disease and the recovery from disease, etc. In general, there are three kinds of epidemic models with age structure: discrete models, continuous models, and stage structure models.

In order to understand epidemic models more easily, we first introduce the age-structured population model.

Population growth model with discrete age structure

We partition the maximum interval in which individuals survive into n equal subintervals, and partition the duration started at time t_0 by the same length as that of the age subinterval as well. Let $N_{ij}(i = 1, 2, 3, \ldots, n, j = 1, 2, 3, \ldots)$ denote the number of individuals whose age is in ith age subinterval at jth time subinterval; let p_i denote the probability that the individual at ith age subinterval still survives at $(i+1)$th age subinterval, that is, $N_{i+1,j+1} = p_i N_{ij}$; and let B_i denote the number of newborn by an individual at ith age subinterval, then the population growth model with discrete age structure is the following:

$$
\begin{cases}
N_{1,j+1} = B_1 N_{1j} + B_2 N_{2j} + B_3 N_{3j} + \cdots + B_n N_{nj} \,, \\
N_{2,j+1} = p_1 N_{1j} \,, \\
\quad\vdots \\
N_{n,j+1} = p_{n-1} N_{n-1,j} \,.
\end{cases}
$$

The discrete system above can be re-written as the following vector difference equation

$$
\mathbf{N}_{j+1} = \mathbf{A}\mathbf{N}_j \,, \tag{19}
$$

where

$$
\mathbf{N}_j = \begin{pmatrix} N_{1j} \\ N_{2j} \\ \vdots \\ N_{nj} \end{pmatrix}, \qquad
\mathbf{A} = \begin{pmatrix}
B_1 & B_2 & B_3 & \cdots & B_{n-1} & B_n \\
p_1 & 0 & 0 & 0 & 0 & 0 \\
0 & p_2 & 0 & 0 & 0 & 0 \\
\vdots & \vdots & \vdots & \vdots & \vdots & \vdots \\
0 & 0 & 0 & \cdots & p_{n-1} & 0
\end{pmatrix}.
$$

Thus, equation (19) is a population growth model with discrete age structure, which is called the Leslie matrix model.

Population growth model with continuously distributed age structure

When the number of individuals is very large and two generations can coexist, this population may be thought to be continuously distributed in age.

Let $f(a,t)\delta\,\mathrm{da}$ denote the number of individuals whose age is between a and $a+\delta\,\mathrm{da}$ at time t, $\gamma(a-\delta\,\mathrm{da})$ the death probability of individuals whose age is between $a-\delta\,\mathrm{da}$ and a in unit time, then we have

$$
f(a-\delta\,\mathrm{da},t) - f(a,t+\delta\,\mathrm{dt}) = \gamma(a-\delta\,\mathrm{da})f(a-\delta\,\mathrm{da},t)\delta\,\mathrm{dt} \,,
$$

where $\delta\,\mathrm{da} = \delta\,\mathrm{dt}$. Taylor expansion of both sides above yields

$$
\frac{\partial f}{\partial t} + \frac{\partial f}{\partial a} + \gamma(a)f = 0 \,.
$$

Let $B(a)\delta\,\mathrm{da}$ denote the mean number of offsprings born by an individual with age between a and $a+\delta\,\mathrm{da}$. Note that $f(0,t)\delta\,\mathrm{da}$ is the number of all the newborn of the population at time t, then we have the boundary condition:

$$
f(0,t)\delta\,\mathrm{da} = \int_0^{+\infty} B(a)f(a,t)\delta\,\mathrm{da} \,,
$$

where $f(a,t)$ is called the distributed function of age density (or age distribution function). From the inference above, we have the equations

$$
\begin{cases}
\frac{\partial f}{\partial t} + \frac{\partial f}{\partial a} + \gamma(a)f = 0 \,, \\
f(0,t)\delta\,\mathrm{da} = \int_0^{+\infty} B(a)f(a,t)\delta\,\mathrm{da} \,, \\
f(a,0) = f_0(a) \,,
\end{cases}
$$

where the last equation is the initial condition.

In the following, we introduce epidemic models with age structure.

Epidemic models with continuous age structure

Many results about epidemic models with continuous age structure have been obtained (Busenberg et al. 1988, 1991; Capasso, V. 1993; Castillo-Chavez et al. 2002; Dietz and Schenzle 1985 Hoppensteadt 1974; Iannelli et al. 1992; Iannelli et al. 1999; Langlais 1995; Li et al. 2001, 2003; Liu et al. 2002; Miiller 1998; Pazy 1983; Zhou 1999; Zhou et al. 2002; Zhu and Zhou 2003). The idea of modelling is the same as that in Sect. 2.2, but all individuals in every compartment are of continuous age distribution. For example, in an SIS model with continuous age structure, the total population is divided into the susceptible compartment and the infected compartment. Let $s(a,t)$ and $i(a,t)$ denote their age distributions at time t, respectively, and assume that the disease transmits only between the same age group. According to the ideas of constructing age-structured population growth models and epidemic compartment models, an SIS model with continuous age structure is given as follows:

$$\begin{cases} \frac{\partial s(a,t)}{\partial a} + \frac{\partial s(a,t)}{\partial t} = -\mu(a)s(a,t) - k_0 i(a,t)s(a,t) + \gamma(a)i(a,t)\,, \\ \frac{\partial i(a,t)}{\partial a} + \frac{\partial i(a,t)}{\partial t} = k_0 i(a,t)s(a,t) - (\gamma(a) + \mu(a))i(a,t)\,, \\ s(0,t)\delta\,\mathrm{d}a = \int_0^A \beta(a)[s(a,t) + (1-q)i(a,t)]\delta\,\mathrm{d}a\,, \\ i(0,t) = \int_0^A q\beta(a)i(a,t)\delta\,\mathrm{d}a\,, \\ s(a,0) = s_0(a),\ i(a,0) = i_0(a)\,, \end{cases} \tag{20}$$

where $\mu(a)$ and $\beta(a)$ are the natural death rate and the birth rate of individuals of age a, $\gamma(a)$ is the rate of recovery from infection at age a, A is the length of maximum survival period of individuals, $k_0(a)$ is the transmission coefficient of the infective of age a, and q is the fraction in which the infectives transmit disease vertically. Using the characteristic method and comparison theorem, Busenberg et al. (1988) proved the following results under some common hypotheses.

Theorem 10. *(Busenberg et al. 1988)   Let $R_0 = q \int_0^A \beta(a)\,\mathrm{e}^{\int_0^a \alpha(\sigma)\delta\,\mathrm{d}\sigma}\delta\,\mathrm{d}a$. The disease-free equilibrium solution $(s^0(a,t),\, i^0(a,t)) = (p_\infty(a), 0)$ is globally asymptotically stable if $R_0 < 1$ and unstable if $R_0 > 1$; The unique endemic equilibrium solution $(s^*(a), i^*(a))$ is globally asymptotically stable if $R_0 > 1$, where $p_\infty(a)$ is the age distribution of total population in the steady state, i. e., $p_\infty(a) = s(a,t) + i(a,t)$ for any $t \geq 0$; $\alpha(\sigma) = -\mu(\sigma) - \gamma(\sigma) + R_0(\sigma)p_\infty(\sigma)$.*

In general, disease can also be transmitted between different age groups. Thus, the term $k_0(a)i(a,t)$ in (20), which reflects the force of the infectivity, should be replaced by term $k_1(a)\int_0^A k_2(a')i(a',t)\delta\,\mathrm{d}a'$, which is the sum of infective force of all the infections to the susceptibles of age a.

For some diseases, if the course of disease is longer and the infectivity may have different courses, then besides the age structure we should also consider the structure with the course of disease. Let c denote the course of disease, then the distributed function with age and course of disease should be denoted by $f(t, a, c)$, and the dimension of the corresponding model will increase and the structure will become more complicated. A few results can be found in references (Hoppensteadt 1974; Zhou et al. 2002; Zhou 1999).

Epidemic models with discrete age structure

Since the unit time of the collection of data about epidemic transmission is usually in days or months, the parameters of the models with discrete age structure can be handled and computed more easily and more conveniently than those with continuous age structure, and these models can sometimes show richer dynamic behaviors. Still, some common methods used for continuous system (such as derivation and integral operation) can not be applied to the discrete system, and so the theoretical analysis of the discrete system will be more difficult. Hence, the results about epidemic models with discrete age structure are few. In order to show the method of constructing models with discrete age structure, we give an SIS model with vertical transmission and death due to disease as follows.

Partition equally the maximum age interval $[0, A]$ into $(m + 1)$ subintervals, and let $\beta_k \lambda_j$ denote the adequate contact rate in which an infected individual whose age belongs to the interval $\left[\frac{kA}{m+1}, \frac{(k+1)A}{m+1}\right]$ $(k = 0, 1, 2, \ldots, m)$ contacts adequately the individuals with age in the interval $\left[\frac{jA}{m+1}, \frac{(j+1)A}{m+1}\right]$ $(j = 0, 1, 2, \ldots, m)$, γ_i the recovery rate of the infective with age in the interval $\left[\frac{jA}{m+1}, \frac{(j+1)A}{m+1}\right]$, d_j and b_j the natural death rate constant and birth rate constant respectively, and $p_j = 1 - d_j$. Thus, according to the ideas of constructing population models with discrete age structure and the epidemic compartment model, we form an SIS model with discrete age structure as follows:

$$
\begin{cases}
S_0(t + 1) = \sum_{j=0}^{m} b_j [S_j(t) + I_j(t)]\,, \\
I_0(t + 1) = 0\,, \\
S_{j+1} = p_j S_j(t) - \lambda_j \sum_{k=0}^{m} \beta_k I_k(t) \frac{S_j(t)}{N_j(t)} + \gamma_j I_j(t)\,, \quad j = 0, 1, 2, \ldots, m - 1\,, \\
I_{j+1} = p_j I_j(t) + \lambda_j \sum_{k=0}^{m} \beta_k I_k(t) \frac{S_j(t)}{N_j(t)} - \gamma_j I_j(t)\,, \quad j = 0, 1, 2, \ldots, m - 1\,, \\
S_j(0) = S_{j0} \geq 0\,, \quad I_j(0) = I_{j0} \geq 0, S_{j0} + I_{j0} = N_j\,, \quad j = 0, 1, 2, \ldots, m\,.
\end{cases}
$$

Allen et al. (1991, 1998) obtained some results for epidemic models with discrete age structure.

Epidemic models with stage structure

In the realistic world, the birth, death, and the infective rate of individuals usually depend on their physiological stage. Thus, investigating the model with stage structure (such as infant, childhood, youth, old age) is significant.

The results (Xiao *et al.* 2002; Xiao and Chen 2002; Xiao and Chen 2003; Lu *et al.* 2003; Zhou et al. 2003) in this fileld are few. In the following, we again introduce an SIS model to show the modelling idea.

We now consider only two stages: larva and adult, and assume that the disease transmits only between larvae. Let $x_1(t)$ denote the number of the susceptibles of the larvae at time t, $x_2(t)$ the number of adults at time t, $y(t)$ the number of the infected larvae of infants at time t, τ the mature period, a_1 the birth rate constant, r the natural death rate constant, b the rate constantfor recovery from disease, and c the coefficient of density dependence of the adults.

Since the mature period of the larvae is τ, the number of transfers out of the larva class at time t is the number of the newborn $a_1 x_2(t-\tau)$ at time $t-\tau$ multiplied by the probability $e^{-r\tau}$ of these newborn who survive until time t. Thus, an SIS model with stage structure and bilinear incidence can be given as follows (Xiao and Chen 2003):

$$\begin{cases} x_1'(t) = a_1 x_2(t) - a_1 e^{-r\tau} x_2(t-\tau) - rx_1(t) - \beta x_1(t)y(t) + by(t)\,, \\ y'(t) = \beta x_1(t)y(t) - by(t) - (r+\alpha)y(t)\,, \\ x_2'(t) = a_1 e^{-r\tau} x_2(t-\tau) - cx_2^2(t)\,, \end{cases} \tag{21}$$

where the term $cx_2^2(t)$ in the last equation of (21) is the density dependence of the adults.

Xiao and Chen (2003) investigated the model (21), obtained the conditions which determine whether the disease dies out or persists, and compared their results with those obtained by the model without stage structure.

2.3.4 Epidemic model with impulses

Impulses can describe a sudden phenomenon which happens in the process of continuous change, such as the reproduction of some algae being seasonal, and vaccinations being done at fixed times of the year. Thus, it is more realistic to describe the epidemic models with these factors by impulsive differential equations.

Concepts of impulsive differential equations

In general, differential equations with impulses happening at fixed times take the following forms:

$$\begin{cases} x'(t) = f(t,x)\,, \quad t \neq \tau_k, k = 1,2,\dots \\ \Delta x_k = I_k(x(\tau_k))\,, \quad t = \tau_k\,, \\ x(t_0) = x_0\,, \end{cases} \tag{22}$$

where $f \in C[R \times R^n, R^n]$ satisfies the Lipschitz condition, $t_0 < \tau_1 < \tau_2 < \cdots$, $I_k \in C[R^n, R^n]\,, \Delta x_k = x(\tau_k^+) - x(\tau_k), x_0 \in R^+, x(\tau_k^+) = \lim\limits_{h \to 0^+} x(\tau_k + h)$.

$x(t)$ is called a **solution** of (22), if it satisfies

1. $x'(t) = f(t, x(t)), t \in [\tau_k, \tau_{k+1})$;
2. $\Delta x_k = x(\tau_k^+) - x(\tau_k)$ for $t = \tau_k$, that is, $x(\tau_k^-) = x(\tau_k)$ and $x(\tau_k^+) = x(\tau_k) + \Delta x_k$.

Since impulsive differential equations are non-automatic, they have no equilibrium. When $\Delta \tau_k = \tau_k - \tau_{k-1}$ is a constant, the existence and stability of the periodic solution with period $\Delta \tau_k$ are often of interest. For further comprehension with respect to impulsive differential equations, see the related references (Lakshmikantham et al. 2003; Bainov and Simeonov 1995; Guo et al. 1995).

Epidemic models with impulsive birth

The study of epidemic models with impulses has started only recently, and related results are scarce (D'Onofri 2002; Jin 2001; Roberts and Kao 1998; Shulgin et al. 1998; Stone et al. 2000; Tang to appear). In the following, we introduce an SIR model with impulsive birth.

Let b denote the birth rate constant , d the natural death rate constant $r = b - d$, and K the carrying capacity of the environment. Assume that there is no vertical transmission and disease-related death, and $\Delta \tau_k = 1$. Note that impulsive birth and the density-dependent term affecting the birth should appear in the impulsive conditions, so the SIS model with impulsive birth is the following:

$$\begin{cases} N'(t) = -dN(t) , & t \neq k , \quad k = 1, 2, 3, \ldots \\ S'(t) = -dS(t) - \beta S(t)I(t) + \gamma I(t) , & t \neq k , \\ I'(t) = \beta S(t)I(t) - (\gamma + d)I(t) , & t \neq k , \\ N(t^+) = \left[1 + b - \frac{rN(t)}{K}\right] N(t) , & t = k , \\ S(t^+) = S(t) + \left[b - \frac{rN(t)}{K}\right] N(t) , & t = k , \\ I(t^+) = I(t) , & t = k . \end{cases}$$

Since $N = S + I$, we only need to discuss the following equations:

$$\begin{cases} N'(t) = -dN(t) , & t \neq k , \quad k = 1, 2, 3, \ldots \\ I'(t) = \beta(N(t) - I(t))I(t) - (\gamma + d)I(t) , & t \neq k , \\ N(t^+) = \left[1 + b - \frac{rN(t)}{K}\right] N(t) , & t = k , \\ I(t^+) = I(t) , & t = k . \end{cases} \tag{23}$$

Theorem 11. *(Han 2002) For model (23), there is always the disease-free periodic solution $(N_1^*(t), 0)$; there exists also the endemic periodic solution $(N_2^*(t), I_2^*(t))$ when $\int_0^1 A(t)\, dt > 0$, where $A(t) = \beta N_1^*(t) - (\gamma + d)$.*

Epidemic models with impulsive vaccination

Assume the fraction p of the susceptibles is vaccinated at time $t = k(k = 0, 1, 2, \ldots)$ and enters into the removed compartment. Then, we have the SIR epidemic model with impulsive vaccination as follows

$$\begin{cases} S'(t) = bK - bS(t) - \beta S(t)I(t)\,, & t \neq k\,, \quad k = 0, 1, 2, \ldots \\ I'(t) = \beta S(t)I(t) - (\gamma + b + d)I(t)\,, & t \neq k\,, \\ R'(t) = \gamma I(t) - bR(t)\,, & t \neq k\,, \\ S(t^+) = (1 - p)S(t)\,, & t = k\,, \\ I(t^+) = I(t)\,, & t = k\,, \\ R(t^+) = R(t) + pS(t)\,, & t = k\,. \end{cases} \qquad (24)$$

For the model (24), Jin and Ma (to appear) obtained the following results by means of Liapunov function and impulsive differential inequalities.

Theorem 12. *(Jin and Ma to appear) Model (24) has always the disease-free periodic solution $(S^*(t), 0, R^*(t))$ with period 1, and it is globally asymptotically stable when $\sigma < 1$, where*

$$S^*(t) = K - \frac{Kpe^{-bt}}{1 - (1-p)e^{-b}}\,,$$

$$R^*(t) = \frac{Kpe^{-bt}}{1 - (1-p)e^{-b}}\,,$$

$$\sigma = \frac{\beta K}{\gamma + b + \alpha}\left[1 - \frac{p(e^b - 1)}{b(e^b - 1 + p)}\right]\,.$$

2.3.5 Epidemic models with multiple groups

Some diseases may be transmitted between multiple interactive populations, or multiple sub-populations of a population. In models constructed for these cases, the number of variables is increased, such that the structure of the corresponding model is complex, and that analysis becomes difficult, so that some new dynamic behaviors can be found. We introduce some modelling ideas as follows.

DI SIA model with different infectivity

In this section, we introduce an epidemic model with different infectivity (DI). Since the differently infected individuals may have different infectivity and different rate of recovery (removed) from a disease, we may partition the infected compartment into n sub-compartments, denoted by $I_i(i = 1, 2, \ldots, n)$, and we let A be the removed compartment in which all the individuals have terminal illness and have no infectivity due to quarantine (for example, the HIV infectives enter into the AIDS period). Assume that all the infectives in compartment I_i can come into contact with the susceptibles, that the infectives in the different sub-compartment I_i have different adequate contact and recovery rates, and that they do not die out due to disease. Thus, we have a DI SIA model with bilinear incidence (Ma, Liu and Li 2003)

$$\begin{cases} S' = \mu S^0 - \mu S - \sum_{i=1}^{n} \beta_i S_i I_i \,, \\[2mm] I_i' = p_i \sum_{i=1}^{n} \beta_i S_i I_i - (\mu + \gamma_i)I \,, \quad i = 1, 2, \ldots, n \,, \\[2mm] A' = \sum_{j=1}^{n} \gamma_j I_j - (\mu + \alpha) \,, \end{cases} \tag{25}$$

where μS^0 denotes a constant input flow, μ the natural death rate constant, α the disease-related death rate constant, γ_i the rate constant of transfer from I_i to A, and β_i the adequate contact number of the infective I_i. p_i is the probability in which the infected individuals enter the compartment I_i, $\sum_{i=1}^{n} p_i = 1$.

Ma et al. (2003) wrote the first $(n+1)$ equations of (25) as the following equations of vector form:

$$\begin{cases} S' = \mu(S^0 - S) - B^{\mathrm{T}} I S \,, \\ I' = S B^{\mathrm{T}} I P - D I \,, \end{cases}$$

where $I = (I_1, I_2, \ldots, I_n)^{\mathrm{T}}, B = (\beta_1, \beta_2, \ldots, \beta_n)^{\mathrm{T}}, D = \mathrm{diag}(\mu + \gamma_1, \mu + \gamma_2, \ldots, \mu + \gamma_n), P = (p_1, p_2, \ldots, p_n)^{\mathrm{T}}$, and T denotes the transposition. Then, it was obtained that

Theorem 13. *(Ma, Liu and Li 2003) Let $R_0 = S^0 B^T D^{-1} P$. The disease-free equilibrium is globally asymptotically stable if $R_0 < 1$ and unstable if $R_0 > 1$; the unique endemic equilibrium is globally asymptotically stable if $R_0 > 1$.*

For Theorem 13, the global stability of the disease-free equilibrium is proved by using the Liapunov function $V = (D^{-1}B)^{\mathrm{T}} I$. To prove the global stability of the endemic equilibrium, the variable translations $S = S^*(1 + x)$, $I_i = I_i^*(1 + y_i)(i = 1, 2, \ldots, n)$ are first made, then the Liapunov function $V = \frac{x^2}{2} + \sum_{i=1}^{n} \frac{\beta_i I_i^*}{\mu + \gamma_i}[y_i - \ln(1 + y_i)]$ is used. At the same time, it is easy to see that R_0 is the basic reproduction number of (25).

DS SIA model with different susceptibility

In this section, we assume that the infected compartment I is homogeneous but the susceptible compartment S is divided into n sub-compartments $S_i(i = 1, 2, \ldots, n)$ according to their susceptibilities, that the input rate of S_i is μS_i^0, and then we have a DS SIA model with different susceptibility and standard incidence

$$\begin{cases} S_i' = \mu(S^0 - S_i) - \frac{\beta k_i S_i I}{N}\,, \\ I' = \sum\limits_{j=1} n\frac{\beta k_j S_j I}{N} - (\mu + \gamma)I\,, \\ A' = \gamma I - (\mu + \alpha)\,, \end{cases} \tag{26}$$

where k_i reflects the susceptibility of susceptible individuals in sub-compartment S_i, and other parameters are the same as those in the previous section. Since the individuals in A do not come into contact with the susceptible individuals, $N = \sum\limits_{i=1}^{n} S_i + I$.

Castillo-Chavez et al. (1996) found the basic reproduction number of (26)

$$R_0 = \frac{\beta \sum\limits_{i=1}^{n} k_i S_i^0}{(\mu + \gamma) \sum\limits_{i=1}^{n} S_i^0}\,,$$

and proved that the disease-free equilibrium is locally asymptotically stable if $R_0 < 1$ and unstable if $R_0 > 1$, and that the disease persists if $R_0 > 1$.

Li et al. (2003) investigated a more complex model, which includes different infectivity and different susceptibility and crossing infections, finding the basic reproduction number, and obtaining some conditions assuring the local and global stability of the disease-free equilibrium and the endemic equilibrium.

Epidemic models with different populations

Though Anderson and May (1986) have incorporated the spread of infective disease into predator-prey models in 1986, the study of disease transmission within interactive populations has started only in recent years. For the investigation of combining epidemic dynamics with population biology, the results obtained are still poor so far.

Anderson and May (1986) assume that the disease transmits only within prey species, and that the incidence is bilinear, the model discussed being

$$\begin{cases} H' = rX - (b + \alpha)Y - c[(1 - f)X + Y]P - bX\,, \\ Y' = \beta XY - (b + \alpha)Y - cYP\,, \\ P' = \delta HP - dP\,, \end{cases} \tag{27}$$

where H is the sum of the number of individuals in the prey species, X the number of the susceptible individuals in the prey species, Y the number of the infective individuals in the prey species, $H = X + Y$, P the number of individuals in the predator species, r the birth rate constant of the prey species, b the natural death rate constant of the prey species, α the disease-related death rate constant of the infectives in the prey species, δ reflects the ability of reproduction of the predator from the prey caught, d the natural death rate constant of the predator, c the catching ability of the predator, and f reflects the difference between catching the susceptible prey and the infected prey. They found that a disease may result in the existence of stable periodic oscillation of two species, which implies that the model (27) has a stable limit cycle.

Some epidemic models of interactive species were discussed (Venturino 1995; Xiao and Chen 2001b; Bowers and Begon 1991; Begon *et al.* 1992; Begon and Bowers 1995; Han et al. 2001). Han et al. (2001) investigated four predator-prey models with infectious disease. Han et al. (2003) analyzed four other SIS and SIRS epidemic models of two competitive species with bilinear or standard incidence and crossing infection, obtaining some complete results where the SIS model with standard incidence is the following

$$
\begin{cases}
S_1' = (b_1 - \frac{a_1 r_1 N_1}{K_1})N_1 - [d_1 + (1 - a_1)\frac{r_1 N_1}{K_1}]S_1 - mN_2 S_1 \\
\quad - \frac{S_1}{N_1}(\beta_{11} I_1 + \beta_{12} I_2) + \gamma_1 I_1 , \\
I_1' = \frac{S_1}{N_1}(\beta_{11} I_1 + \beta_{12} I_2) - \gamma_1 I_1 - [d_1 + (1 - a_1)\frac{r_1 N_1}{K_1}]I_1 - mN_2 I_1 , \\
N_1' = [r_1(1 - \frac{N_1}{K_1}) - mN_2]N_1 , \\
S_2' = (b_2 - \frac{a_2 r_2 N_2}{K_2})N_2 - [d_2 + (1 - a_2)\frac{r_2 N_2}{K_2}]S_2 - nN_1 S_2 \\
\quad - \frac{S_2}{N_2}(\beta_{21} I_1 + \beta_{22} I_2) + \gamma_2 I_2 , \\
I_2' = \frac{S_2}{N_2}(\beta_{21} I_1 + \beta_{22} I_2) - \gamma_2 I_2 - [d_2 + (1 - a_2)\frac{r_2 N_2}{K_2}]I_2 - nN_1 I_2 , \\
N_2' = [r_2(1 - \frac{N_2}{K_2}) - nN_1]N_2 , \\
r_i = b_i - d_i > 0 , \quad i = 1, 2 \\
0 \leq a_i \leq 1 , \quad i = 1, 2
\end{cases}
\tag{28}
$$

The explanation of parameters in (28) is omitted. Since $N_i = S_i + I_i (i = 1, 2)$, (28) can be simplified as follows:

$$
\begin{cases}
I_1' = \frac{N_1 - I_1}{N_1}(\beta_{11} I_1 + \beta_{12} I_2) - \gamma_1 I_1 \\
\quad - [d_1 + (1 - a_1)\frac{r_1 N_1}{K_1}]I_1 - mN_2 I_1 , \\
N_1' = [r_1(1 - \frac{N_1}{K_1}) - mN_2]N_1 , \\
I_2' = \frac{N_2 - I_2}{N_2}(\beta_{21} I_1 + \beta_{22} I_2) - \gamma_2 I_2 \\
\quad - [d_2 + (1 - a_2)\frac{r_2 N_2}{K_2}]I_2 - nN_1 I_2 , \\
N_2' = [r_2(1 - \frac{N_2}{K_2}) - nN_1]N_2 , \\
N_i \geq I_i \geq 0 , \quad i = 1, 2 \\
0 \leq a_i \leq 1 , \quad i = 1, 2
\end{cases}
\tag{29}
$$

The model has six boundary equilibria and one positive equilibrium,and the attractive region of all feasible equilibria has been determined. The results obtained show that, under certain conditions, the disease can die out eventually by cutting off the inter-infections between two species or decreasing the inter-transmission coefficients between two species to a fixed value.

2.3.6 Epidemic models with migration

The models in the previous sections do not include the diffusion or migration of individuals in space, and suppose that the distribution of individuals is uniform. In fact, with the migration of individuals, the influence of individual diffusion on the spread of disease should not be neglected. Here, we introduce two types of diffusions into the epidemic models.

First, we consider the continuous diffusion of individuals in the corresponding compartment. This needs to add the diffusion to the corresponding ordinary differential equations. For example, the SIR model with diffusion corresponding to model (4) is

$$\begin{cases} \frac{\partial S}{\partial t} = \Delta S + bK - \beta SI - bS\,, \\ \frac{\partial I}{\partial t} = \Delta I + \beta SI - (b+\gamma)S\,, \\ \frac{\partial R}{\partial t} = \Delta R + \gamma I - bR\,, \end{cases} \tag{30}$$

where $S = S(t,x,y,z), I = I(t,x,y,z)$ and $R = R(t,x,y,z)$ denote the numbers of the susceptibles, the infectives, and the removed individuals at time t and point (x,y,z), respectively; $\Delta S = \frac{\partial^2 S}{\partial x^2} + \frac{\partial^2 S}{\partial y^2} + \frac{\partial^2 S}{\partial z^2}, \Delta I = \frac{\partial^2 I}{\partial x^2} + \frac{\partial^2 I}{\partial y^2} + \frac{\partial^2 I}{\partial z^2}$ and $\Delta R = \frac{\partial^2 R}{\partial x^2} + \frac{\partial^2 R}{\partial y^2} + \frac{\partial^2 R}{\partial z^2}$ are the diffusion terms of the susceptibles, the infectives, and the removed individuals at time t and point (x,y,z), respectively. This model is a quasi-linear partial differential system. Model (30), with some boundary conditions, constitutes an SIR epidemic model with diffusion in space.

Second, we consider the migration of individuals among the different patches (or regions). Though Hethcote (1976) established an epidemic model with migration between two patches in 1976, studies dealing with this aspect are rare. Brauer et al. (2001) discussed an epidemic model with migration of the infectives. Recently, Wang (2002) considered an SIS model with migration among n patches. If there is no population migration among patches, that is, the patches are isolated, according to the structure of population dynamics proposed by Cooke et al. (1999), the epidemic model in ith ($i = 1, 2, \ldots, n$) patch is given by

$$\begin{cases} S_i' = B_i(N_i)N_i - \mu_i S_i - \beta_i S_i I_i + \gamma_i I_i\,, \\ I_i' = \beta_i S_i I_i - (\gamma_i + \mu_i)I_i\,, \end{cases}$$

where the birth rate in ith patch $B_i(N_i)$ for $N_i > 0$ satisfies the following common hypothesis:

$$B_i(N_i) > 0\,, \quad B_i(N_i) \in C^1(0, +\infty)\,, \quad B_i'(N_i) < 0\,, \quad \text{and} \quad \mu_i > B_i(+\infty)\,.$$

If n patches are connected with each other, i.e., the individuals between any two patches can migrate, then the SIS epidemic model with migration among n patches is the following:

$$\begin{cases} S_i' = B_i(N_i)N_i - \mu_i S_i - \beta_i S_i I_i + \gamma_i I_i + \sum\limits_{j=1}^{n} a_{ij} S_j\,, \\ I_i' = \beta_i S_i I_i - (\gamma_i + \mu_i)I_i + \sum\limits_{j=1}^{n} b_{ij} I_j\,, \end{cases} \tag{31}$$

where a_{ii} and b_{ii} $(a_{ii} \leq 0, b_{ii} \leq 0)$ denote the migration rates of the susceptibles and the infectives from the ith patch to other patches, respectively; a_{ij} and b_{ij} $(a_{ij} \geq 0, b_{ij} \geq 0)$ denote the immigration rates of the susceptibles and the infectives from the jth patch to ith patch, respectively. Model (31) assumes that the disease is not fatal, and the death and birth of individuals in the process of migration are neglected. Since the individuals migrating from the ith patch will move dispersedly to the other $(n-1)$ patches, we have

$$-a_{ii} = \sum_{\substack{j=1 \\ j \neq i}}^{n} a_{ji}, \qquad -b_{ii} = \sum_{\substack{j=1 \\ j \neq i}}^{n} b_{ji}.$$

Under the assumptions that the matrices (a_{ij}) and (b_{ij}) are all in-reducible, by means of related theory of matrix, Wang (2002) obtained the conditions of local and global stability of the disease-free equilibrium, and the conditions under which the disease persists in all patches. Particularly, for the case of two patches, the conditions about the disease-free equilibrium obtained by Wang (2002) show the following: when the basic reproduction number R_{12}, which is found when regarding two patches as one patch, is greater than one, the disease persists in two patches; when $R_{12} < 1$ and $R_{12} - \Phi_{12} > 1$ (where Φ_{12} denotes one number minus the other number, the first number is the product of the number of new patients infected by an infected individual within the average infective course in one patch and that in another patch, the second number is the product of the number of migrated patients within the average infective course in one patch and that in another patch), the disease still persists; when $R_{12} < 1$ and $R_{12} - \Phi_{12} < 1$, the disease dies out in two patches. The formulation above indicates: $R_{12} < 1$ can not ensure the extinction of the disease, and the condition $R_{12} - \Phi_{12} < 1$ is also added. This result shows that migration among patches can affect the spread of disease.

Besides those research directions mentioned above, there are some other research directions of epidemic dynamics, such as: using non-autonomous models, where the coefficients of the epidemic model are time dependent, including periodic coefficients and more general time-dependent coefficients (Lu and Chen 1998;Ma 2002); combining epidemic dynamics with eco-toxicology to investigate the effect of pollution on the spread of disease in a polluted environment (Wang and Ma 2004); combining epidemic dynamics with molecular biology to investigate the interaction among viruses, cells and medicines inside the body (De Boer et al. 1998; Lou et al. 2004a, 2004b; Perelson et al. 1993; Wang et al. to appear; Wang et al. 2004); combining epidemic dynamics with optimal control to investigate the control strategy of epidemics (He 2000); considering stochastic factors to investigate the stochastic dynamics of epidemics (Jing 1990); and using some special disease to construct and investigate a specific model (Feng and Castillo-Chavez 2000; Hethcote and Yorke 1984). Because of limited space, we can not discuss these one by one. In the following, we only introduce the modelling and investigation of SARS according to the real situation existing on mainland China in 2003.

2.3.7 Epidemic models for SARS in China

SARS (Severe Acute Respiratory Syndrome) is a newly acute infective disease with high fatality. This infection first appeared and was transmitted within China in November 2002, and spread rapidly to 31 countries within 6 months. In June 2003, the cumulative number of diagnosed SARS cases had reached 8454, of which 793 died in the whole world (WHO; MHC). In China, 5327 cases were diagnosed, and 343 cases died (MHC).

Since SARS had never been recorded before, it was not diagnosed correctly and promptly, and there have been no effective drugs or vaccines for it so far. Therefore, investigating its spread patterns and development tendency, and analyzing the influence of the quarantine and control measures on its spread are significant. In the initial period of onset of SARS, some researchers (Chowell et al. 2003; Donnelly et al. 2003; Lipsitch et al. 2003; Riley et al. 2003) studied its spread rule and predicted its development according to the data published at that time. Based on the data available for China, Zhang et al. (2004) and Zhou et al. (2004) established some continuous and discrete dynamic models, and discovered some transmission features of SARS in China, which matched the real situation quite well.

Continuous model for SARS in China

The difficulties we met in the modelling of SARS are the following: (1) because SARS is a new disease, the infectious probability is unknown, and whether the individuals in the exposed compartment have infectivity is not sure; (2) how to construct the model such that it fits the situation in China? Especially, how to account for those effective control measures carried out by the government, such as various kinds of quarantine, and how to obtain data for those parameters which are difficult to quantify, for example, the intensity of the quarantine?

Based on the general principles of modelling of epidemics, and the special case of the prevention and control measures in China, Zhang et al. (2004) divided the whole population into two related blocks: the free block in which the individuals may move freely, and the isolated block in which the individuals were isolated and could not contact the individuals in the free block. Further, the free block was divided into four compartments: the susceptible compartment (S), the exposed compartment (E), the infectious compartment (I), and the removed compartment (R); the isolated block was divided into three compartments: the quarantined compartment (Q), the diagnosed compartment (D), and the health-care worker compartment (H).

The susceptible compartment (S) consisted of individuals susceptible to the SARS virus; the individuals in the exposed compartment (E) were exposed to the SARS virus, but in the latent period (these were asymptomatic but possibly infective); the individuals in the infectious compartment (I) showed definitive symptoms, and had strong infectivity, but had not yet been isolated; the individuals in the removed compartment (R) were those who had recovered from SARS, with full immunity against re-infection. The individuals in the quarantined compartment (Q) were either individuals carrying the SARS virus (but not yet diagnosed) or individuals without the SARS virus but misdiagnosed as possible SARS patients; the individuals in the diagnosed compartment (D) were carriers of SARS virus and had been diagnosed; the health-care worker compartment (H) consisted of those who were health-care workers with high susceptibility (since SARS is not known well), and were quarantined due to working with the individuals in the isolated block.

To control and prevent the spread of SARS, the Ministry of Health of China (MHC) decreed the Clinic Diagnostic Standard of SARS, and imposed strict measures of quarantine at that time. According to these measures, any individual who came into contact with a diagnosed patient with SARS directly or indirectly, or had clinical symptoms similar to those of SARS, such as fever, chills, muscular pain, and shortness of breath, would be quarantined as a possible SARS patient. These measures played a very important role in controlling the spread of SARS in China. Inevitably, many individuals were misdiagnosed as SARS suspected, and hence were mistakenly put in the Q-compartment due the to lack of a fast and effective SARS diagnostic test. According to the relations among all compartments, the transfer diagram of SARS should be Fig. 2.7.

Let $S(t), E(t), I(t), R(t), Q(t), D(t)$, and $H(t)$ denote the number of individuals in the compartments S, E, I, R, Q, D, and H at time t, respectively.

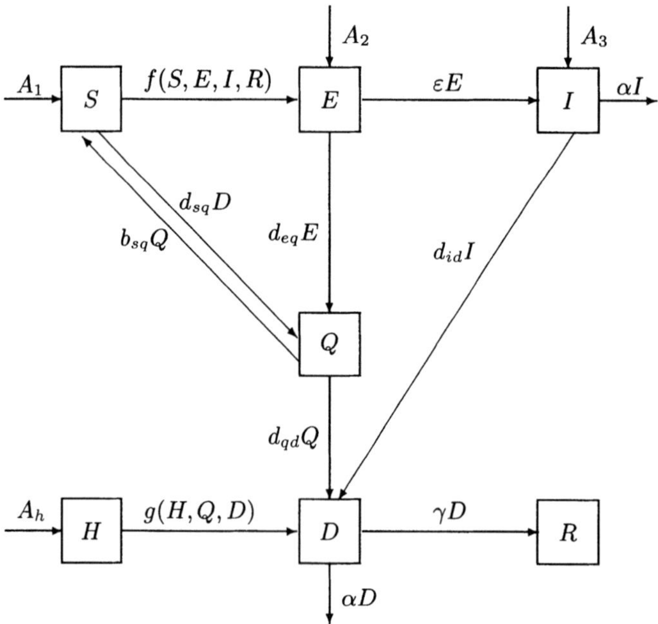

Fig. 2.7. Transfer diagram for the SARS model in China

Thus, corresponding to the transfer diagram in Fig. 2.7, we have the compartment model of SARS as follows:

$$\begin{cases} S' = A_1 - f(S, E, I, R) - d_{sq}D + b_{sq}Q \,, \\ E' = A_2 + f(S, E, I, R) - \varepsilon E - d_{eq}E \,, \\ I' = A_3 + \varepsilon E - d_{id}I - \alpha I \,, \\ Q' = d_{eq}E + d_{sq}D - b_{sq}Q - d_{qd}Q \,, \\ D' = g(H, Q, D) + d_{qd}Q + d_{id}I - \alpha D - \gamma D \,, \\ H' = A_h - g(H, Q, D) \,, \\ R' = \gamma D \,, \end{cases} \tag{32}$$

where $F(S, E, I, R)$ and $g(H, Q, D)$ are the incidences in the free block and the isolated block, respectively. The general form of the incidences is $\beta C \frac{S}{N} I$, where β is the probability of transmitting the virus per unit time of effective contact (this measures the toxicity of the virus), and C is the adequate contact number of a patient with other individuals (this reflects the strength of control and prevention against SARS). Let C_E and C_I denote the contact rates, and let β_E and β_I denote the probabilities of transmission of exposed individuals and infective individuals in the free block, respectively. Then, the incidence in the free block is given by

$$f(S, E, I, R) = (\beta_E C_E E + \beta_I C_I I) \frac{S}{S + E + I + R} \,.$$

Here, to be on the safe side, we suppose that the individuals in the exposed compartment have a small infectivity, that is, $0 < \beta_E \ll \beta_I$. Similarly, we can get

$$g(H, Q, D) = (\beta_Q C_Q Q + \beta_D C_D D) \frac{H}{H + Q + D} \,.$$

For the sake of simplicity, let $k_1 = \frac{\beta_E C_E}{\beta_I C_I}$ denote the ratio of the infectivity between an individual in the E-compartment and an individual in the I-compartment, then we rewrite the incidence terms $f(S, E, I, R)$ as

$$f(S, E, I, R) = \beta(t)(k_1 E + I)$$

where $\beta(t) = \frac{\beta_I C_I S}{S + E + I + R}$ represents the infectious rate.

We took one day as unit time, and assumed that the average latent period is 5 days (WHO; MHC). From the statistical data published by MHC (Rao and Xu 2003), each day 80% of the diagnosed SARS cases come from the Q-compartment, and 20% come from the I-compartment. So, we let

$$\epsilon = \frac{1}{5} \times \frac{20}{100} \,, \quad d_{eq} = \frac{1}{5} \times \frac{80}{100}$$

Since the average number of days from entering the I-compartment to moving to the D-compartment is 3 days, $d_{id} = \frac{1}{3}$. On the other hand, if we assumed that the average transition times from the Q-compartment to the D-compartment, and from the Q-compartment to the S-compartment (those are misdiagnosed) are 3 days and 10 days, respectively, then by denoting the number of removed from quarantines to susceptibles and that of diagnosed from quarantines by q_s and q_d respectively, based on the daily reported data from MHC, we get the ratio of non-infected individuals in the Q-compartment, $\frac{q_s}{q_s+q_d} = 0.6341$. Thus,

$$d_{qd} = (1 - 0.6341) \times \frac{1}{3}, \quad b_{sq} = 0.6341 \times \frac{1}{10}.$$

Since the period of recovery for an SARS patient is about 30 days and the statistical analysis from the MHC shows that the ratio of the daily number of new SARS suspected cases to the daily number of new SARS diagnosed cases is 1.3:1, then

$$\gamma = \frac{1}{30}, \quad d_{sd} = 1.3 \times 0.6341 \times \frac{1}{30}.$$

Finally, since the probability of SARS-related death is 14%, $\alpha = \frac{1}{30} \times 0.14$.

The determination of the incidences in the free block and the isolated block is the key to analyzing the SARS model (32). This is difficult because of the poor understanding of the SARS virus toxicities and the difficulty in quantifying these quarantines. Nevertheless, significant amounts of data have been collected during the course of SARS outbreak in China after the middle of April 2003. Here, we use the back-tracking method to estimate the adequate contact rate.

Let $\hat{f}(t)$ denote the number of new diagnosed SARS cases (reported by MHC) minus the number of new diagnosed SARS cases in the H-compartment at time t. $F(t) := f(S(t), E(t), I(t), R(t))$, the new infectives at time t (tth day) in the free block should be $\hat{f}(t+8)$ because the average number of days from exposure to the SARS virus to the definite diagnosis is 8 days, with the first 5 days in the E-compartment (with low infectivity) and the last 3 days in the I-compartment (with high infectivity). Therefore,

$$\beta(t) = \frac{\beta_I C_I S}{S + E + I + R} = \frac{F(t)}{I(t) + k_1 E(t)} = \frac{\hat{f}(t+8)}{\sum\limits_{j=0}^{2} \hat{f}(t+j) + k_1 \sum\limits_{j=3}^{7} \hat{f}(t+j)}.$$

Analogously, we can obtain the incidence in the isolated block.

By regression analysis of the data published by WHC, we obtain

$$\beta(t) = 0.002 + 0.249\, e^{-0.1303t}.$$

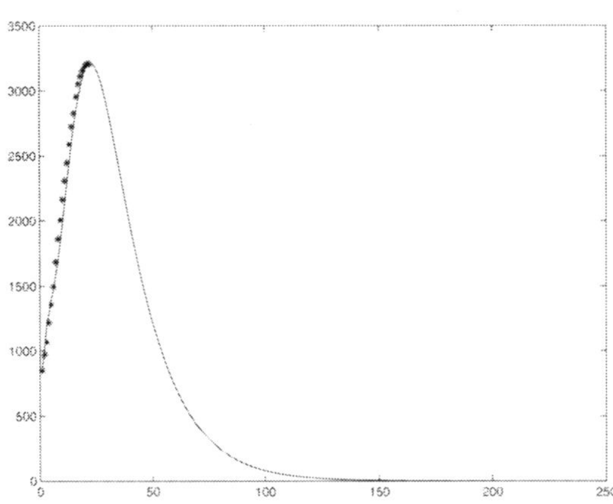

Fig. 2.8. The simulated curve (continuous) and the reported (by MHC, marked in stars) daily number of SARS patients in hospitals

Based on the above approach, Zhang et al. (2004) carried out some numerical simulations to validate the model (32), and discussed the effectiveness of control measures, and to assess the influence of certain measures on the spread of SARS, by varying some parameters to gauge the effectiveness of different control measures. All numerical simulations started on April 21 of 2003, that is, the origin of the time axis (horizontal) corresponds to April 21

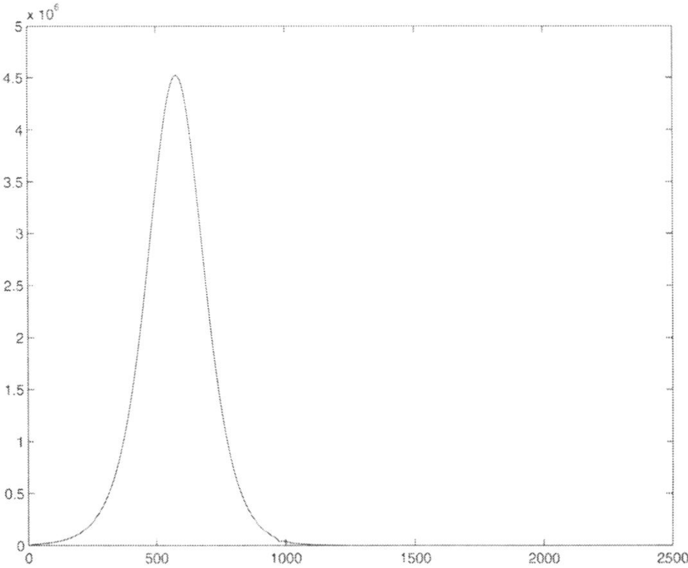

Fig. 2.9. The prediction for the transmission pattern in China without control measures after April of 2003. The outbreak peaks at the end of October of 2004, with over 4.5 million individuals infected, though the SARS toxicity declines exponentially

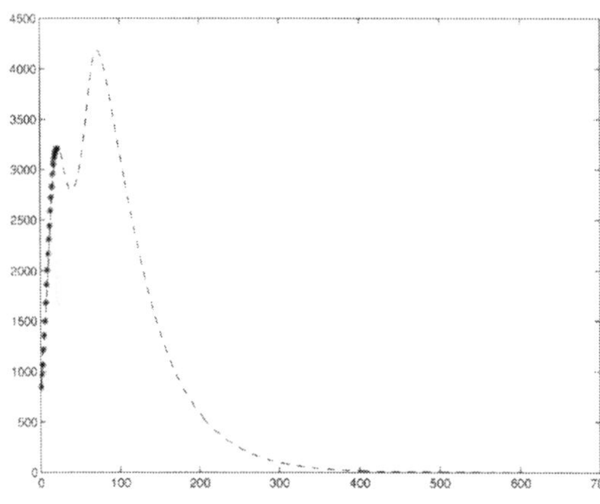

Fig. 2.10. The pattern of SARS transmission if the prevention and control measures were to have been relaxed from May 19 of 2003; though the toxicity of SARS virus naturally declines at a rate of 0.01 per unit time, there would be a second outbreak, with a maximum number of SARS patients higher than that of the first outbreak

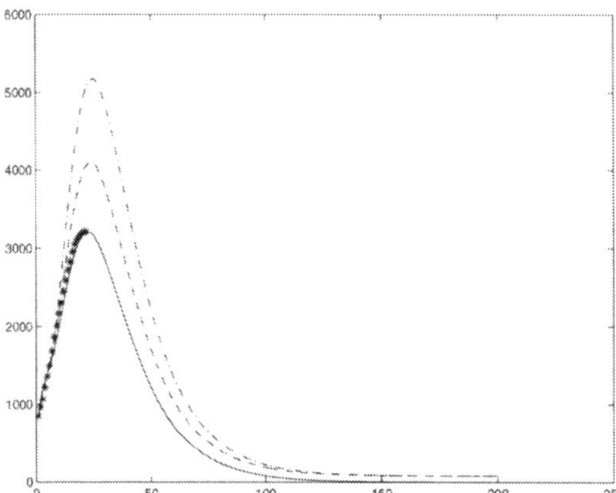

Fig. 2.11. The influence of the slow quarantine speed: the top and bottom lines show the number of daily SARS patients when the infectives stay in the I-compartment for 2 and 1 more days, respectively. The peak would be postponed by 4 or 2 days, but the peak numbers would be much higher

of 2003. Figure 2.8 shows the simulated curve of the daily number of SARS patients in hospitals in reality. Figure 2.9 shows the case with no control measures after April 21 of 2003. Figure 2.10 shows the case under which the prevention and control measures were relaxed from May 19 of 2003 onwards. Figure 2.11 shows the influence of the slow quarantine speed.

From the simulations above, we consider that the rapid decrease of the SARS patients can be attributed to the high successful quarantine rate and timely implementation of the quarantine measures, and indeed all of the prevention and control measures implemented in China are very necessary and effective.

Discrete model for SARS in China

Zhou et al. (2004) followed the same idea of modelling the transmission of SARS in China as that presented above, and proposed a discrete model for SARS in China. However, the susceptible compartment, which includes the S-compartment and the H-compartment, was omitted in this case, because the number of susceptible individuals was extremely large compared with the number of individuals in other compartments, and some SARS patient can not contact all the population.

Zhou et al. (2004) made the following assumptions: the new infected exposed is proportional to the sum $kE(t) + I(t)$; individuals in the E-compartment move to the I-compartment and Q-compartment at the rate constant ϵ and λ, respectively; individuals in the Q-compartment move to the D-compartment at the rate constant σ; individuals in the I-compartment move to the D-compartment at the rate θ; individuals in the D-compartment move to the R-compartment at rate constant γ; d and α are the natural death rate constant and the SARS-induced death rate constant, respectively. Then, the model proposed is

$$\begin{cases}
E(t+1) = E(t) + \beta(t)[kE(t) + I(t)] - (d + \epsilon + \lambda)E(t)\,, \\
I(t+1) = I(t) + \epsilon E(t) - (d + \alpha + \theta)I(t)\,, \\
Q(t+1) = Q(t) + \lambda E(t) - (d + \sigma)Q(t)\,, \\
D(t+1) = D(t) + \theta I(t) + \sigma Q(t) - (d + \alpha + \gamma)D(t)\,, \\
R(t+1) = R(t) + \gamma D(t) - dR(t)\,.
\end{cases} \tag{33}$$

Similarly to the methods used by Zhang et al. (2004), Zhou et al. (2004) determined the parameters in model (33).

Zhang et al. (2004) and Zhou et al. (2004) varied some parameters to analyze the effectiveness of different control and quarantine measures. These new parameters corresponded to the situation when the quarantine measures in the free block were relaxed or when the quarantine time of SARS patients was postponed. The purpose of the introduction of these new parameters was to demonstrate the second outbreak with a maximum number of daily SARS patients and a delayed peak time. They obtained the basic reproduction number, and their results agree quite well with the developing situation of SARS in China.

References

1. Allen L. J. S.,and D. B. Thrasher (1998), The effects of vaccination in an age-dependent model for varicella and herpes zoster, IEEE Transactions on Automatic Control, **43**, 779.
2. Allen, L. J. S., M. A. Jones and C. E. Martin (1991), A Discrete-time model with vaccination for a measles epidemic, Math. Biosci., **105**, 111.
3. Anderson, R. and R. May (1982), *Population Biology of infectious diseases*, Spring-Verlag, Berlin, Heidelberg, New York.
4. Anderson, R. M. and R. M. May (1986), The invasion, persistence, and spread of infectious diseases within animal and plant communities, Phil. Trans. R. Soc., **B314**, 533.
5. Baily, N. T. J. (1975), *The Mathematical theory of infectious disease*, 2nd edn, Hafner, New York.

6. Bainov,D. D. and P. S. Simeonov (1995), *Impulsive Differential Equations: Asymptotic Properties of the solutions*, World Scientific, New Jersey London.

7. Begon, M. and R. G. Bowers (1995), Host-host-pathogen models and microbial pest control: the effect of host self-regulation, J. Theor. Biol., **169**, 275.

8. Begon, M., R. G. Bowers and N. Kadianakis, et al. (1992), Disease and community structure: the importance of host-regulation in a host-host-pathogen model, Am. Nat., **139**,1131.

9. Bowers, R. G. and M. Begon (1991), A host-pathogen model with free living infective stage, applicable to microbial pest control, J. Theor. Biol., **148**, 305.

10. Brauer, F and C. Castillo-Chavez (2001), Mathematical Models in Population Biology and Epidemiology. In Marsden, J. E., L. Sirovich and M. Golubitsky, *Texts in Applied Mathematics*, vol 40, Springer-Verlag, New York Berlin Heidelberg.

11. Brauer, F. and P. van den Driessche (2001), Models for transmission of disease with immigration of infectives, Math. Biosci., **171**, 143.

12. Busenberg, S. and P. van den Driessche (1990), Analysis of a disease transmission model in a population with varying size, J. Math. Biol., **28**, 257.

13. Busenberg, S., K. Cooke and M. Iannelli (1988), Endemic, thresholds and stability in a case of age-structured epidemics, SIAM J. Appl. Math., **48**, 1379.

14. Busenberg, S., M. Iannelli and H. R. Thieme (1991), Global behavior of an age-structed epidemic model, SIAM J. Math. Anal., **22**, 1065.

15. Capasso, V. (1993), Mathematical Structures of Epidemic systems, In C. Castillo-Chavez, *Lecture Notes in Biomathematics*, vol 97, Springer-Verlag, Heidelberg.

16. Castillo-Chavez, C., W. Z. Huang and J. Li (1996), Competitive exclusion in gonorrhea models and other sexually transmitted diseases, SIAM J. Appl. Math., **56**, 494.

17. Castillo-Chavez, C., S. Blower and P. van den Driessche (2002), Mathematical Approaches for Emerging Infectious Diseases. In C. Castillo-Chavez, *The IMA volumes in mathematics and its applications*, vol 125, Springer-Verlag, New York Berlin Heidelberg.

18. Chowell, G., P. W. Fenimore, M. A. Castillo-Garsow and C. Castillo-Chavez (2003), SARS outbreaks in Ontario, Hong Kong, and Singapore: the role of diagnosis and isolation as a control mechanism, Los Alamos Unclassified Report LA-UR-03-2653.

19. Cooke, K., P. van den Driessche and X. Zou (1999), Interaction of maturation delay and nonlinear birth in population and epidemic models, J. Math. Biol., **39**, 332.

20. De Boer, R. J. and A. S. Perelson (1998), Target cell limited and immune control models of HIV infection: a comparison, J. Theor. Biol., **190**, 201.

21. Dietz, K. (1982), Overall population patterns in the transmission cycle of infectious disease agents, In C. Castillo-Chavez, *Population Biology of infectious disease*, vol 56, Springer, New York.

22. Dietz, K. and D. Schenzle (1985), Proportionate mixing models for age-dependent infection transmission, J. Math. Biol., **22**, 117.

23. Donnelly, C. A., et al. (2003), Epidemiological determination of spread of causal agent of severe acute respiratory syndrome in Hong Kong. The Lancent, Publish online May 7, http://image.thelacent.com/extras/03 art 4453 web.psf.

24. D'Onofri, A.(2002), Stability properties of pulse vaccination strategy in SEIR epidemic model, Math. Biosci., **179**, 57.
25. Dushoff, J., W. Huang and C. Castillo-Chavez (1998), Backward bifurcation and catastrophe in simple models of fatal disease, J. Math. Biol., **36**, 227.
26. Fan, M. and M. Y. Li, K. Wang (2001), Global stability of an SEIS epidemic model with recruitment and a varying total population size, Math. Biosci., **170**, 199.
27. Feng, Z. and C. Castillo-Chavez (2000), A model for tuberculosis with exogenous reinfection, Theoret. Popul. Biol., **57**, 235.
28. Feng, Z. and H. R. Thieme (2003a), Endemic models with arbitrarily distributed periods of infection, I: general theory, SIAM J. Appl. Math., **61**, 803.
29. Feng, Z. and H. R. Thieme (2003b), Endemic models with arbitrarily distributed periods of infection, II: fast disease dynamics and permanent recovery, SIAM J. Appl. Math.,**61**, 983.
30. Guo, D. J., J. X. Sun and Z. L. Zhao (1995), *Methods of Nonlinear Scientific Functionals*, Shandong Scientific Press, Jinan.
31. Hadeler, K. and P. van den Driessche (1997), Backward bifurcation in epidemic control, Math. Biosci., **146**, 15.
32. Hamer, W. H.(1906), Epidemic disease in England, Lancet, **1**, 733.
33. Han, L. T.(2002), Study on epidemic models of two interaction species. DR Thesis, Xi'an Jiaotong University, Xi'an, China.
34. Han, L. T., Z. E. Ma and T. Shi (2003), An SIRS epidemic model of two competitive species, Mathl. Comput. Modelling, **37**, 87.
35. Han, L. T., Z. E. Ma and H. M. Hethcote (2001), Four predator prey model with infectious diseases, Mathl. Comput. Modelling, **34**, 849.
36. He, Z. R. (2000), Optimal control of population dynamical systems with age-structure. DR Thesis, Xi'an Jiaotong University, Xi'an, China
37. Heesterbeek, J. A. P. and J. A. J. Metz (1993), The saturating contact rate in marrige and epidemic models, J. Math. Biol., **31**, 529.
38. Hethcote, H. (1976), Qualitative analysis of communicable disease models, Math. Biosci., **28**, 335.
39. Hethcote, H. W. and P. van den Driessche (1995), An SIS epidemic model with variable population size and a delay, J. Math. Biol.,**34**, 177.
40. Hethcote, H. W. and J. A. Yorke (1984), Gonorrhea transmission dynamics and control. In H. W. Hethcote, *Lecture Notes Biomathematics*, vol 56, Springer-Verlag, Berlin-Heidelberg-New York.
41. Hethcote, H., Z. E. Ma, and S. Liao (2002), Effects of quarantine in six endemic models for infectious diseases, Math. Biosci., **180**, 141.
42. Hoppensteadt, F. (1974), An age dependent epidemic model, J. Franklin Inst., **197**, 325.
43. Iannelli, M., F. A. Miller and A. Pugliese (1992), Analytic and numerical results for the age-structured SIS epidemic model with mixed inter-intracohort transmission, SIAM J. Math. Anal., **23**, 662.
44. Iannelli, M., M. Y. Kim and J. Park (1999), Asymptotic behavior for an SIS epidemic model and its approximation, Nonlinear Analysis: ATM, **35**, 797.
45. Jin, Z. (2001), The study for ecological and epidemic models influenced by impulses. DR Thesis, Xi'an Jiaotong University, Xi'an, China.
46. Jin, Z. and Z. E. Ma, Epidemic models with continuous and impulse vaccination, to appear.

47. Jing, Z. J. (1990), Mathematics and AIDS, Practice and Understanding of Mathematics, **20**, 47.
48. Kermack, W. O. and A. G. McKendrick (1932), Contributions to the mathematical theory of epidemics, Proc. Roy. Soc., **138**, 55.
49. Kermack, W. O. and A. G. McKendrick (1927), Contributions to the mathematical theory of epidemics, Proc. Roy. Soc., **115**, 700.
50. Lakshmikantham, V., D. D. Bainov and P. S. Simeonov (1989), *Theory of Impulsive Differential Equations*, World Scientific Press, London.
51. Langlais, M. (1995), A mathematical analysis of the SIS intra-cohort model with age-structure. In C. Castillo-Chavez, *Mathematical Population Dynamics*, vol 1, Wuerz Publishing Ltd., Winning, Canada, pp 103–117.
52. Li, J. Q. and Z. E. Ma (2003), Qualitative analysis of an epidemic model with vaccination, Annals of Differential Equations, **19**, 34.
53. Li, J. Q. and Z. E. Ma (2004), Global analysis of SIS epidemic models with variable total population size, Mathl. Comput. Modelling,**39**, 1231.
54. Li, J. Q. and Z. E. Ma (2004), Stability analysis for SIS epidemic models with vaccination and constant population size, Discrete and Continuous Dynamical Systems: Series B, **4**, 637.
55. Li, J. Q. and Z. E. Ma(2002), Qualitative analyses of SIS epidemic model with vaccination and varying total population size, Mathl. Comput. Modelling, **35**, 1235.
56. Li, J. Q. and Z. E. Ma, Global stability of two epidemic models with vaccination, to appear
57. Li, J. Q., Z. E. Ma and Y. C. Zhou, Global analysis of SIS epidemic model with a simple vaccination and multiple endemic equilibria, to appear.
58. Li, J., Z. E. Ma and S. P. Blythe and C. Castillo-Chavez (2003), On coexistence of pathogens in a class of sexually transmitted diseases models, J. Math. Biol., **47**, 547.
59. Li, M. Y. and J. S. Muldowney (1995), Global stability for the SEIR in epidemiology, Math. Biosci., **125**, 155.
60. Li, M. Y. and L. Wang, Global stability in some SEIR epidemic models, to appear.
61. Li, M. Y. and J. R. Graef, L. Wang, et al. (1999), Global dynamics of a SEIR model with varying total population size, Math. Biosci., **160**, 191.
62. Li, M. Y. and J. S. Muldowney (1996), A geometric approach to the global stability problems, SIAM J. Math. Anal., **27**, 1070.
63. Li, M. Y., H. L. Smith and L. Wang (2001), Global dynamics of an SEIR epidemic model with vertical transmission, SIAM J. Appl. Math., **62**, 58.
64. Li, X. Z., G. Gupur and G. T. Zhu (2001), Threshold and stability results for an age-structured SEIR epidemic model, Comp. Math. Appl., **42**, 883.
65. Li, X. Z.. and G. Gupur and G. T. Zhu (2003), Mathematical Theory of Age-Structured Epidemic Dynamics. Research Information Ltd., Hertfordshire
66. Lipsitch, M., et al. (2003), Transmission dynamics and control of severe acute respiratory syndrome, Published online May 23, 2003; 10.1126/science.1086616 (Science Express Reports).
67. Liu, Han-Wu, J. Xu and Q. M. Xu (2002), SIS epidemic models with continuous vaccination and age structure, Chinese J. Engin. Math., **19**, 25.
68. Liu, Wei-min., H. W. Hethcote and S. A. Levin (1987), Dynamical behavior of epidemiological model with nonlinear incidence rates, J. Math. Biol., **25**, 359.

69. Liu, Wei-min., S. A. Levin and Y. Iwasa (1986), Influence of nonlinear incidence rates upon the behavior of SIRS epidemiological models, J. Math. Biol., **23**, 187.

70. Lizana, M. and J. Rivero (1996), Multiparametric bifurcation for a model in epidemiology, J. Math. Biol., **35**, 21.

71. Lou, J., Z. E. Ma and J. Q. Li, et al. (2004a), The impact of the CD8+ cell non-cytotoxic antiviral response (CNAR) and cytotoxic T lymphocyte (CTL) activity in a cell-to-cell spread model for HIV-1 that includes a time delay, Journal of Biological Systems, **12**, 73.

72. Lou, J., Z. E. Ma and Y. M. Shao, et al. (2004b), Modelling the interaction of T cells, antigen presenting cells and HIV in vivo, to appear in Comput. Math. Appl.

73. Lu, Z. H. and L. S. Chen (1998), Threshold theory of a nonautonomous SIS-infectious diseases model with logistic species growth. In *Advanced topics in biomathematics, Proc. of ICMB'97' Hangzhou, China*, World Scientific Press, Singapore.

74. Lu, Z. H., S. J. Gao and L. S. Chen (2003), Analysis of an SI epidemic model with nonlinear transmission and stage structure, Acta Math. Scientia, **4**, 440.

75. Ma, J. L.(2002), Study of threshold of nonautonomous infectious diseases models. MA Thesis, Xi'an Jiaotong University, Xi'an, China.

76. Ma, W. B., T. Hara, Y. Takeuchi (2002), et al., Permanence of an SIR epidemic model with distributed time delays, Tohoku Math., **54**, 365.

77. Ma, Z. E. and Y. C. Zhou (2001), *Methods of Qualitativeness and Stability for Ordinary Differential Equations*, Science Press, Beijing.

78. Ma, Z. E., J. P. Liu and J. Li (2003), Stability analysis for differential infectivity epidemic models, Nonlinear Analysis: RWA, **4**, 841.

79. Ma, Z. E., Y. C. Zhou, W. D. Wang, Z. Jin (2004), *Mathematical Modelling and Study of Epidemic Dynamics*, Science Press, Beijing.

80. Mena-Lorca, J. and H. W. Hethcote (1992), Dynamic models of infectious diseases as regulators of population sizes, J. Math. Boil., **30**, 693.

81. MHC, Available from http://168.160.244.167/sarsmap.

82. MHC, Available from
http://www.moh.gov.cn/zhgl/xgxx/fzzsjs/1200304290124.htm.

83. MHC, Available from
http://www.moh.gov.cn/zhgl/xgxx/fzzsjs/1200306030008.htm.

84. Miiller, J. (1998), Optimal vaccination patterns in age-structured populations, SIAM J. Appl. Math., **59**, 222.

85. Muldowney, J. S. (1990), Compound matrices and ordinary differential equations, Rocky Mount. J. Math., **20**, 857.

86. Pazy, A.(1983), *Semigroups of Linear Operators and Applications to Partial differential Equations*, Springer-Verlag, Berlin Heidelberg.

87. Perelson, A. S., D. E. Kirschner and R. J. De Boer (1993), Dynamics of HIV infection of CD$^+$ T cells, Math. Biosci., **114**, 81.

88. R. V. Culshaw and S. Ruan (2003), A mathematical model of cell-to-cell spread of HIV-1 that includes a time delay, J. Math. Biol., **46**, 189.

89. Rao, K. and D. Xu 2003, The talk on CCTV on May 9.

90. Riley, S., et al. (2003), Transmission dynamics of the etiological agent of SARS in Hong Kong: impact of public health interventions, Published online May 23, 2003; 10.1126/science.1086478 (Science Express Reports).

91. Roberts, M. G. and R. R. Kao (1998), The dynamics of an infectious disease in a population with birth pulses, Math. Biosci., **149**, 23.

92. Ross, R. (1911), *The prevention of Malaria*, 2nd edn, Murray, London.

93. Ruan, S. G. and W. D. Wang (2003), Dynamic behavior of an epidemic model with a nonlinear incidence rate, J. Diff. Eqns., **18**, 135.

94. Shulgin, B., L. Stone and Z. Agur (1998), Pulse vaccination strategy in the SIR epidemic model, Bull. Math. Biol., **60**, 1123.

95. Stone, L., B. Shulgin and Z. Agur(2000), Theoretical examination of the pulse vaccination policy in the SIR epidemic model, Mathl. Comput. Modeling, **31**, 207.

96. Tang, S. Y., Y. N. Xiao and L. S. Chen, The dynamics of SIS epidemic models with pulse vaccination, to appear.

97. Van den Driessche P. and J. Watmough (2000), A simple SIS epidemic model with a backward bifurcation, J. Math. Biol., **40**, 525.

98. Venturino, E. (1995), Epidemics in predator-prey models: Disease in the prey. In C. Castillo-Chavez, *Mathematical population dynamics : Analysis of heterogeneity*, vol 1, Wuerz Publishing, Canada.

99. Wang, F. and Z. E. Ma (2004), Persistence and periodic orbits for SIS model in a polluted environment, to appear in Comput. Math. Appl.

100. Wang, F., Z. Ma and Y. M. Shao (2004), A competition model of HIV with recombination effect, to appear on Mathl. Comput. Modelling.

101. Wang, F., Z. Ma and Y. M. Shao, Recombination HIV model with time delay, to appear.

102. Wang, W. D. (2002a), Global Behavior of an SEIRS epidemic model time delays, Appl. Math. Lett., **15**, 423.

103. Wang, W. D. (2002b), Stability and bifurcation of epidemic mathematical models, DR Thesis, Xi'an Jiaotong University, Xi'an, China.

104. Wang, W. D. and Z. E. Ma(2002), Global dynamics of an epidemic model with time delay. Nonlinear Analysis: Real Word Applications, **3**, 365.

105. WHO and CDC, Available from http://www.moh.gov.cn/was40/detail?record =14&channelid=8085 &searchword=%B7%C7%B5%E4%D2%DF%C3%E7.

106. WHO, Available from http://www.who.int/csr/sars/country/2003_06_13/en.

107. World Health Organization Report (1995).

108. Wu, L-I. and Z. Feng (2000), Homoclinic bifurcation in an SIQR model for childhood diseases, J. Diff. Eqnt., **168**, 150.

109. Xiao, Y. N. and L. S. Chen (2003), On an SIS epidemic model with stage structure, J. Sys. Sci. and complexity, **16**, 275.

110. Xiao, Y. N. and L.. S. Chen (2001a), Analysis of a SIS epidemic model with stage structure and a delay, Communications in Nonlinear Sci. and Numer. Simul., **1**, 35.

111. Xiao, Y. N. and L..S. Chen (2001b), Modelling and analysis of a predator-prey model with disease in the prey, Math. Biosci., **170**, 59.

112. Xiao, Y. N., L. S. Chen and F. ven den Bosch (2002), Dynamical behavior for a stage-structured SIR infectious disease model, Nonlinear Analysis: Real Word Applications, **3**, 175.

113. Xiao,Y. N. and L. S. Chen (2002), A SIS epidemic model with stage structure and a delay, Acta Math. Appl. Sinica, English Series, **18**, 607.

114. Yuan, S. and Z. Ma (2001), Global stability and Hopf bifurcation of an SIS epidemic model with time delays, J. System Sci. Complexity, **14**, 327.

115. Yuan, S. and Z. Ma (2002), Study on an epidemic model with time variant delay, J. System Sci. Complexity, **14**, 299.

116. Yuan, S., L.. Han and Z. Ma (2003a), Analysis of an SIS epidemiologic model with variable population size and a delay, Appl. Math. J. Chinese Univ. Ser.B, **16**, 9.

117. Yuan, S., Z. Ma and Z. Jin (2003b), Persistence and periodic solution on nonautonomous SIS model with delays, Acta Appl. Math. Sinica, English Series, **19**, 167.

118. Zhang, J. and Z. E. Ma (2003), Global dynamics of an SEIR epidemic model with saturating contact rate, Math. Biosci., **185**, 15.

119. Zhang, J. and Z. E. Ma, Global dynamics of an SEIR epidemic model with population dependent incidence, to appear.

120. Zhang, J., J. Lou, Z. E. Ma and J. H. Wu (2004), A compartmental model for the analysis of SARS transmission patterns and outbreak control measures in China, to appear in Appl. Math. Computations.

121. Zhou, Y. C. (1999), Epidemic models with age-structure and the infective courses and their asymptotical behavior. DR Thesis, Xi'an Jiaotong University, Xi'an, China.

122. Zhou, Y. C., B. J. Song and Z. E. Ma (2002), The global stability analysis for emerging and reemerging infectious diseases, In: C. Castillo-Chavez *The IMA volumes in mathematics and its applications*, vol 125, Springer-Verlag, New York Berlin Heidelberg.

123. Zhou, Y. C., Z. E. Ma and F. Brauer (2004), A discrete epidemic model for SARS transmission and control in China, to appear on Mathl. Comput. Modelling.

124. Zhou, Y., M. Z. Lou and J. Lou (2003), Study of an type of epidemic model with age stage, Chinese J. Engin. Math., **20**, 135.

125. Zhu, L. Y. and Y. C. Zhou (2003), Global stability of the positive steady solution of epidemic models with vaccination and age structure, J. BioMath., **18**, 27.

3

Delayed SIR Epidemic Models for Vector Diseases *

Yasuhiro Takeuchi and Wanbiao Ma

Summary. The purpose of the chapter is to give a survey on the recent researches on SIR models with time delays for epidemics which spread in a human population via a vector. Based on the Hethcote model and Cooke's SIS model with a time delay, we introduce SIR models with time delays and a constant population size. Further, the SIR models are modified in such a way that the death rates for three population classes are different. Finally, the models are revised to assume that the birth rate is not independent of the total population size. For all models, we summarize the known mathematical results on stability of the equilibria and permanence. We also give some open problems and our conjectures on the threshold for an epidemic to occur.

3.1 Introduction

It is well known (see, for example, Hethcote (1976); Anderson and May (1979)) that the spread of a communicable disease involves not only disease-related factors such as infectious agent, mode of transmission, incubation periods, infectious periods, susceptibility, and resistance, but also social, cultural, economic, demographic, and geographic factors. Mathematical models have become important tools in analyzing the spread and control of infectious diseases. The models provide conceptual ideas such as thresholds, basic reproduction numbers, contact numbers and replacement numbers. Communicable disease models involving a directly transmitted viral or bacterial agent in a closed population consisting of **susceptibles** (S), **infectives** (I), and **recovereds** (R) were considered by Kermack and McKendrick (1927). Their model was governed by a nonlinear integral equation from which further leads to the following well known SIR Kermack-McKendrick model without

* The research of this project is partially supported by the National Natural Science Foundation of China (No. 10671011) and the Foundation of USTB for WM and the Ministry, Science and Culture in Japan, under Grand-in-Aid for Scientific Research (A) 13304006 for YT.

vital dynamics (see, for example, Kermack and McKendrick (1927); Thieme (2003); Ruan (2005)):

$$\frac{dS}{dt} = -\beta S(t)I(t),$$
$$\frac{dI}{dt} = \beta S(t)I(t) - \gamma I(t), \tag{1}$$
$$\frac{dR}{dt} = \gamma I(t).$$

In the model, it is assumed that there is no birth and death in the population and that the total numbers of population is constant, i. e., $N(t) = S(t)+I(t)+R(t) = \text{const.}$. The positive constants β and γ are called the daily contact rate and the daily recovery rate, respectively. A complete theoretical analysis on dynamics properties of (1) can be done easily. A simple but very primary SIR epidemic model with vital dynamics is proposed and studied by Hethcote (1976):

$$\frac{dS}{dt} = \mu - \beta S(t)I(t) - \mu S(t),$$
$$\frac{dI}{dt} = \beta S(t)I(t) - \gamma I(t) - \mu I(t), \tag{2}$$
$$\frac{dR}{dt} = \gamma I(t) - \mu R(t),$$

where the total number of population is assumed to be $N(t) = S(t) + I(t) + R(t) = 1$, the positive constant μ is the birth and death rate. For the SIR model (2), the following result is well known:

Theorem 1 (Hethcote 1976). *Let*

$$\sigma = \frac{\beta}{\mu + \gamma}.$$

If $\sigma \leq 1$, then the disease free equilibrium

$$E_0 = (1, 0, 0)$$

is globally asymptotically stable.

If $\sigma > 1$, then the disease free equilibrium E_0 becomes unstable and the only endemic equilibrium

$$E_+ = \left(\frac{1}{\sigma}, \frac{\mu(\sigma - 1)}{\beta}, \frac{\gamma(\sigma - 1)}{\beta}\right)$$

is globally asymptotically stable.

Theorem 1 implies that, for a disease with vital dynamics where recovery gives permanent immunity to the disease, if the infectious contact number σ exceeds one, then the susceptible, infective and recovered fractions eventually approach constant positive endemic values. If the infectious contact number is less than or equal to one, then the infective and recovered fractions eventually approach zero, and hence, the all population member becomes eventually susceptible.

Based on the Kermack-McKendrick model, various epidemic models have been developed in recent decades, such as SI models, SIS models, SIRS models, SIR models, SEIRS models, SEIR models with or without time delays (see, for example, Smith (1983); Liu et al. (1986); Mena-Lorca and Hethcote (1996); Cooke and van den Driessche (1996); Cooke et al. (1999); Hethcote and van den Driessche (2000); Chen et al. (2002); Ma et al.(2004) and the references there in).

The purpose of the chapter is to give a survey on the recent researches on SIR models with time delays for epidemic which spreads in a human population **via a vector.** In the next section, based on Hethcote model (2) and Cooke's SIS model with a time delay, we introduce SIR models with time delays and a constant population size. Further, the SIR models are modified in such a way that the death rates for three population classes are different. In Sect. 3.3, the models are revised to assume that the birth rate is not independent of the total population size. We assume the the number of newborns is proportional to the total population size. In the final section, the number is assumed to saturate when the total population size increases. For all models, we summarize the known mathematical results on stability of the equilibria and permanence. We also give some open problems and our conjectures on the threshold for an epidemic to occur.

3.2 SIR epidemic models with constant birth rate and time delays

3.2.1 Stability analysis of SIR epidemic models with constant population

It is well known that Cooke (1979) proposed an SIS model for epidemic which spreads in a human population **via a vector** (such as mosquito etc). The model considers the case where susceptible individuals (denoted by $S(t)$) receive the infection from an infectious vector, and susceptible vectors receive the infection from infectious individuals. If it is assumed that when a susceptible vector is infected by an infectious person, there is a time $\tau > 0$ during which the infectious agents develop in the vector and it is only after that time that the infected vector becomes itself infectious. Cooke's epidemic models with time delays are stated as follows:

$$\dot{y}(t) = by(t - \tau)[1 - y(t)] - cy(t)\,, \tag{3}$$

where $y(t)$ denotes the infective individuals of population who are infectious at time t. $b > 0$ and $c > 0$ are the contact rate and the recovery rate, respectively. It is assumed in (3) that the vectors population is very large and the total number of human population is constant, i. e., $S(t) + y(t) = 1$, and that at any time t the infectious vector population is simply proportional to the human infectious population at time $t - \tau$. A detailed analysis on the local stability and global stability of the nonnegative equilibrium of (3) was given by Cooke (1979).

Based on the motivations in the papers by Hethcote (1976), Anderson and May (1979), Cooke (1979), Di Liddo (1986), Cushing (1977) and MacDonald (1978), Beretta et al. (1988), and Beretta and Takeuchi (1995, 1997) proposed the following two classes of SIR epidemic models with discrete time delays or distributed time delays:

$$\frac{\mathrm{d}S}{\mathrm{d}t} = \mu - \beta S(t) \int_{-\infty}^{t} F^{(k)}(t - \tau)I(\tau)\,\mathrm{d}\tau - \mu S(t)\,,$$

$$\frac{\mathrm{d}I}{\mathrm{d}t} = \beta S(t) \int_{-\infty}^{t} F^{(k)}(t - \tau)I(\tau)\,\mathrm{d}\tau - \gamma I(t) - \mu I(t)\,, \tag{4}$$

$$\frac{\mathrm{d}R}{\mathrm{d}t} = \gamma I(t) - \mu R(t)\,,$$

$$\frac{\mathrm{d}S}{\mathrm{d}t} = \mu - \beta S(t) \int_{0}^{\infty} f(s)I(t - s)\,\mathrm{d}s - \mu S(t)\,,$$

$$\frac{\mathrm{d}I}{\mathrm{d}t} = \beta S(t) \int_{0}^{\infty} f(s)I(t - s)\,\mathrm{d}s - \gamma I(t) - \mu I(t)\,, \tag{5}$$

$$\frac{\mathrm{d}R}{\mathrm{d}t} = \gamma I(t) - \mu R(t)\,,$$

and

$$\frac{\mathrm{d}S}{\mathrm{d}t} = \mu - \beta S(t)I(t - h) - \mu S(t)\,,$$

$$\frac{\mathrm{d}I}{\mathrm{d}t} = \beta S(t)I(t - h) - \gamma I(t) - \mu I(t)\,, \tag{6}$$

$$\frac{\mathrm{d}R}{\mathrm{d}t} = \gamma I(t) - \mu R(t)\,,$$

where the positive constants μ, β and γ have the same biological meanings as in the model (2). The constant $h \geq 0$ is a time delay. The kernel functions $F^{(k)}(\tau)$ and $f(s)$ are continuous, nonnegative and assumed to satisfy the conditions

$$F^{(k)}(\tau) = \frac{\alpha^k}{(k - 1)!}\tau^{k-1}e^{-\alpha\tau}\,, \quad (k \in N_+, \alpha > 0.)$$

for (4),

$$\int_{0}^{\infty} f(s)\,\mathrm{d}s = 1$$

and

$$\int_0^\infty sf(s)\,\mathrm{d}s < \infty$$

for (5). It is assumed that the total number of populations is constant, i.e.,

$$N(t) = S(t) + I(t) + R(t) = 1\,.$$

Note that models (4)–(6) completely have the same equilibria as system (2), i.e., there always exists the disease free equilibrium $E_0 = (1,0,0)$; if the infectious contact number is larger than one,

$$\sigma = \frac{\beta}{\mu + \gamma} > 1\,,$$

then, there also exists the endemic equilibrium

$$E_+ = \left(\frac{1}{\sigma}, \frac{\mu(\sigma - 1)}{\beta}, \frac{\gamma(\sigma - 1)}{\beta}\right)\,.$$

From the mathematical and biological points of view, it is usually an important problem to give a complete analysis for the local stability and global stability of the disease free equilibrium E_0 and the endemic equilibrium E_+. The models (4) and (5) belong to a class of functional differential equations with infinite time delays, their stability analysis is closely related to the choice of phase spaces. However, with help of the well known linear chain technique, the model (4) can be easily induced into a high-dimensional nonlinear ordinary differential equations, and hence, the global stability of its nonnegative equilibira can be discussed based on the method of Liapunov functions (Beretta, Capasso and Rinaldi (1988)). For the models (5) and (6), the following results are obtained by Beretta and Takeuchi (1995) by constructing proper Liapunov functionals:

Theorem 2 (Beretta and Takeuchi 1995). *For the model (5), if $\sigma < 1$, then the disease free equilibrium $E_0 = (1,0,0)$ is locally asymptotically stable; if $\sigma > 1$, then the disease free equilibrium E_0 becomes unstable, and the endemic equilibrium E_+ is locally asymptotically stable.*

Theorem 3 (Beretta and Takeuchi 1995). *For the model (6), if $\sigma < 1$, then the disease free equilibrium $E_0 = (1,0,0)$ is globally asymptotically stable; if $\sigma > 1$, then the disease free equilibrium E_0 becomes unstable, and the endemic equilibrium E_+ is locally asymptotically stable.*

3.2.2 Stability analysis of SIR epidemic models with varying population

Beretta and Takeuchi (1997) remove some unrealistic assumptions on the parameters in the model (6). It is assumed that the death rates of three classes of the population are different, that is susceptibles, infectives and recovered have $\mu_i (i = 1, 2, 3)$ as their death rates respectively. It is further assumed that the birth rate $b > 0$ is different from the death rate. All newborns are assumed to be susceptible, again. Then we have the following SIR epidemic models with distributed time delays:

$$\frac{\mathrm{d}S}{\mathrm{d}t} = b - \beta S(t) \int_0^h f(s) I(t - s) \, \mathrm{d}s - \mu_1 S(t) \, ,$$

$$\frac{\mathrm{d}I}{\mathrm{d}t} = \beta S(t) \int_0^h f(s) I(t - s) \, \mathrm{d}s - \gamma I(t) - \mu_2 I(t) \, , \qquad (7)$$

$$\frac{\mathrm{d}R}{\mathrm{d}t} = \gamma I(t) - \mu_3 R(t) \, ,$$

where the positive constants β and γ have same biological meanings as in the model (1). The constant $h \geq 0$ is a time delay. The kernel $f(s)$ is continuous, nonnegative and assumed to satisfy the condition

$$\int_0^h f(s) \, \mathrm{d}s = 1 \, .$$

For the models (7), there always exists the disease free equilibrium $E_0 = (S_0, 0, 0)$, where

$$S_0 \equiv \frac{b}{\mu_1} \, .$$

If

$$S_0 > S^* \equiv \frac{\mu_2 + \gamma}{\beta} \, , \qquad (8)$$

there also exists the endemic equilibrium $E_+ = (S^*, I^*, R^*)$, where

$$I^* = \frac{b - \mu_1 S^*}{\beta S^*} \, , \qquad R^* = \frac{\gamma}{\mu_3} I^* \, .$$

For the models (7), the following results are obtained by Beretta and Takeuchi (1997):

Theorem 4 (Beretta and Takeuchi 1997). *If $S_0 < S^*$, then the disease free equilibrium $E_0 = (S_0, 0, 0)$ is globally asymptotically stable; if $S_0 > S^*$, then the disease free equilibrium E_0 becomes unstable, and the endemic equilibrium $E_+ = (S^*, I^*, R^*)$ is locally asymptotically stable. Furthermore, if $S_0 > S^*$, it is possible to construct an explicit region Ω such that, for any initial function belonging to Ω, the solutions $(S(t), I(t), R(t))$ of (7) tends to $E_+ = (S^*, I^*, R^*)$ as t tends infinity.*

Takeuchi, Ma and Beretta (2000) further considered attractivity of the disease free equilibrium E_0 of (7) while $S_0 = S^*$, by using Liapunov-LaSalle invariant principle (Hale, 1977; Kuang, 1993), and had the following result:

Theorem 5 (Takeuchi, Ma and Beretta 2000). *If $S_0 = S^*$, then the disease free equilibrium $E_0 = (S_0, 0, 0)$ is globally attractive.*

Theorem 5 shows that the disease will eventually disappear under any length of time delay h whenever the endemic equilibrium does not exist. For global stability of the endemic equilibrium $E_+ = (S^*, I^*, R^*)$ of (7), based on some inequality techniques, the following results are obtained:

Theorem 6 (Takeuchi, Ma and Beretta 2000). *If*

$$S_0 > S^* ,$$

and there is some $\tilde{S}$ satisfying

$$S^* < \tilde{S} < b/(\mu_2 + \gamma)$$

such that the following conditions hold:

$$(i) \quad h < \min\left\{ \left(2\beta\tilde{S}\right)^{-1}, \frac{\tilde{S} - S^*}{b - \mu_1 S^*} \right\} ;$$

$$(ii) \quad b < \tilde{S}\left[\beta\left(\frac{b}{\mu_2 + \gamma} - \tilde{S} \right) + \mu_1 \right] ,$$

then the endemic equilibrium $E_+ = (S^*, I^*, R^*)$ is globally asymptotically stable.

Theorem 7 (Takeuchi, Ma and Beretta 2000). *If*

$$S_0 > S^* ,$$

and the following conditions hold:

$$(iii) \quad b\beta > (\mu_2 + \gamma)^2 \left(2 - \frac{\mu_1}{\mu_2 + \gamma} + 2\sqrt{1 - \frac{\mu_1}{\mu_2 + \gamma}} \right) ;$$

$$(iv) \quad h < \min\left\{ (2\beta g)^{-1}, \frac{g - S^*}{b - \mu_1 S^*} \right\} ;$$

where

$$g = \frac{1}{2\beta}\left[\frac{b\beta}{\mu_2 + \gamma} + \mu_1 + \sqrt{\left(\frac{b\beta}{\mu_2 + \gamma} + \mu_1 \right)^2 - 4\beta b} \right] ,$$

then the endemic equilibrium $E_+ = (S^*, I^*, R^*)$ is globally asymptotically stable.

In mathematics, Theorems 6–7 give the sufficient conditions to ensure the global asymptotic stability of the endemic equilibrium E_+ whenever it exists. In biology, Theorems 6–7 show that, while E_+ exists, the disease always remains endemic if the time delay h is short enough and the product βb of the contact constant β and the birth rate b is relatively large (or the death rates μ_1 and μ_2, and the recovery rate γ are small enough).

Based on Hethcote's result, i. e., Theorem 1 for the SIR epidemic model (2) without delay, and general properties for delay differential equations, the following natural conjecture was proposed by Takeuchi, Ma and Beretta:

Problem 1 (Takeuchi, Ma and Beretta, 2000). For sufficiently small time delay $h \leq h_0$, condition (8) should imply the global asymptotic stability of the endemic equilibrium E_+, i. e. condition (8) should be the threshold of (7) for an epidemic to occur.

In fact, note that Theorems 4 and 5 for the SIR epidemic model (7) with time delays and numerical simulations given by Ma, Takeuchi, Hara and Beretta (2002), it is strongly suggested that the follwoing more general conclusion may be true:

Problem 2 (Ma, Takeuchi, Hara and Beretta 2002). For any time delay h, condition (8) should imply the global asymptotic stability of the endemic equilibrium E_+.

Unfortunately, Theorems 6 and 7 need more restrictive conditions in order to ensure the global asymptotic stability of the endemic equilibrium E_+.

Under the assumption of permanence of model, Beretta, Hara, Ma and Takeuchi (2001) gave a positive answer for the above Problem 1 for a class of SIR epidemic model with more general time delays by constructing a complicated Liapunov functional.

Consider the following SIR epidemic model with distributed time delays:

$$\frac{dS}{dt} = b - \beta S(t) \int_0^h I(t - s)\,d\eta(s) - \mu_1 S(t)\,,$$

$$\frac{dI}{dt} = \beta S(t) \int_0^h I(t - s)\,d\eta(s) - \gamma I(t) - \mu_2 I(t)\,, \tag{9}$$

$$\frac{dR}{dt} = \gamma I(t) - \mu_3 R(t)\,,$$

where the positive constants $b, \beta, \mu_i (i = 1, 2, 3)$ and γ have same biological meanings as in the model (7). The constant $h \geq 0$ is a time delay. The function $\eta(s)$ is nondecreasing and has bounded variation such that

$$\int_0^h d\eta(s) = \eta(h) - \eta(0) = 1\,.$$

The model (9) has the same the equilibria as the model (7). The following result was proved by Beretta et al.:

Theorem 8 (Beretta et al. 2001). *Suppose that $S_0 > S^*$, and the model (9) is permanent, that is, there are positive constants ν_i and $M_i (i =$*

$1, 2, 3)$ *such that*

$$\nu_1 \leq \liminf_{t \to +\infty} S(t) \leq \limsup_{t \to +\infty} S(t) \leq M_1 \; ;$$

$$\nu_2 \leq \liminf_{t \to +\infty} I(t) \leq \limsup_{t \to +\infty} I(t) \leq M_2 \; ;$$

$$\nu_3 \leq \liminf_{t \to +\infty} R(t) \leq \limsup_{t \to +\infty} R(t) \leq M_3$$

hold. Then the endemic equilibrium $E_+ = (S^*, I^*, R^*)$ of (9) is globally asymptotically stable, if the time delay h is small enough such that

$$H \equiv \int_0^h s \, d\eta(s) < H_0 \, ,$$

where the positive constant H_0 can be explicitly expressed by the parameters in (9) and the constants ν_i and $M_i (i = 1, 2, 3)$.

3.2.3 Permanence of SIR epidemic models with varying population

It is well known that *permanence* of dynamical systems plays a very important role in the studying of population dynamical systems. Based on some known techniques on limit sets of differential dynamical systems developed by Butler et al. (1986); Hale and Wlatman (1989) and Freedman and Ruan (1995), Ma et al. (2002) give a partial answer to Problem 1.

Theorem 9 (Ma et al. 2002). *For any time delay $h \geq 0$, (8) is necessary and sufficient for permanence of (9).*

In biology, Theorem 9 implies that, for any time delay $h \geq 0$, (8) gives the threshold of the model (9) for an endemic to occur.

It should be pointed out here that, in the proof of Theorem 9, the positive constant ν_2 appeared in Theorem 8 is only shown to exist in a theoretical form.

Based on some known results for epidemic models with time delays by Cooke and van den Driessche (1996) and Hethcote and van den Driessche (2000,1995) and Wang (2002) proposed two classes of SEIRS epidemic model with time delays and also discussed global behavior of the equilibria of the models. With help of some analysis techniques developed by Wang (2002), Song et al. (2005, 2006) recently further considered permanence of a class of more general SIR epidemic model with non-constant birth rate and time delays, and gave an explicit expression for the positive constant ν_2, which plays an important role in the application of Theorem 8. We shall give a detailed description in the following section.

3.3 SIR epidemic models with time delays and non-constant birth rate $bN(t)$

Takeuchi and Ma (1999) considered the following delayed SIR epidemic model with density dependent birth rate:

$$\dot{S}(t) = -\beta S(t)I(t - h) - \mu_1 S(t) + bN(t) \, ,$$
$$\dot{I}(t) = \beta S(t)I(t - h) - (\mu_2 + \gamma)I(t) \, , \qquad (10)$$
$$\dot{R}(t) = \gamma I(t) - \mu_3 R(t) \, ,$$

where $S(t) + I(t) + R(t) \equiv N(t)$ denotes the total number of population at time t. The positive constants b, $\mu_1, \mu_2, \mu_3, \beta$ and γ have the same biological meanings as in model (7). The nonnegative constant h is a time delay. Note that, in (10), the birth rate of population is dependent on the total number of population $N(t)$.

For (10), we have the following classification on the existence of its equilibria.

(i)   Equation (10) always has a trivial equilibrium $E_0 = (0,\ 0,\ 0)$.

(ii)   If $b = \mu_1$, then for any $S > 0, E_S = (S, 0, 0)$ is the boundary equilibrium (the disease free equilibrium) of (10).

(iii) If

$$b = \mu_1 = \frac{\mu_3(\mu_2 + \gamma)}{\mu_3 + \gamma} \, ,$$

then for any $I > 0$ and $R > 0$ such that $\gamma I = \mu_3 R$, $E_{SR} = (S^*, I, R)$ is the positive equilibrium (the endemic equilibrium) of (10), where

$$S^* \equiv \frac{\mu_2 + \gamma}{\beta} \, .$$

(iv)   If

$$\mu_1 < b < \frac{\mu_3(\mu_2 + \gamma)}{\mu_3 + \gamma} \, , \qquad (11)$$

then (10) has a unique positive equilibrium $E_+ = (S^*, I^*, R^*)$, where

$$S^* \equiv \frac{\mu_2 + \gamma}{\beta} \, , \quad I^* \equiv \frac{\mu_3(b - \mu_1)S^*}{\beta S^* \mu_3 - b(\mu_3 + \gamma)} \, , \quad R^* \equiv \frac{\gamma}{\mu_3}I^* \, .$$

3.3.1 Global asymptotic properties

For the model (10), it has the following results:

Theorem 10 (Takeuchi and Ma 1999).

(a) If $\mu_1 > b$, then the boundary equilibrium E_0 is globally asymptotically stable.

(b) If $b > \mu_1$, then E_0 is unstable. Further, if

$$b > \frac{\mu_3(\mu_2 + \gamma)}{\mu_3 + \gamma} \,,$$

then, for any solution $(S(t), I(t), R(t))^T$ of (10),

$$\lim_{t \to +\infty} N(t) = \lim_{t \to +\infty} (S(t) + I(t) + R(t)) = +\infty \,.$$

Theorem 11 (Takeuchi and Ma 1999). *If*

$$\mu_1 = b < \frac{\mu_3(\mu_2 + \gamma)}{\mu_3 + \gamma} \,,$$

then, for any solution $(S(t), I(t), R(t))^T$ of (10), there is some constant $c \geq 0$ such that

$$c \leq S^* = (\mu_2 + \gamma)/\beta$$

and

$$\lim_{t \to +\infty} S(t) = c \,, \qquad \lim_{t \to +\infty} I(t) = \lim_{t \to +\infty} R(t) = 0 \,.$$

Moreover, we also have the following

Theorem 12 (Ma and Takeuchi 2004). *If*

$$b > \mu_1$$

and

$$b \geq \frac{\mu_3(\mu_2 + \gamma)}{\mu_3 + \gamma} \,,$$

then, for any solution $(S(t), I(t), R(t))^T$ of (10),

$$\lim_{t \to +\infty} (S(t) + I(t)) = +\infty \,.$$

In [29], the convergence of the positive equilibrium E_+ had also been considered by using Liapunov functionals.

3.3.2 Hopf bifurcation and local stability on E_+

Ma and Takeuchi (2004) further considered local asymptotic stability of the positive equilibrium E_+ and Hopf bifurcation of (10) based on well-known Hopf bifurcation theorem (see, for example, Hale (1977) and Kuang (1993)). We have the following:

Theorem 13 (Ma and Takeuchi 2004). *If (11) holds, then there exist a positive constant sequence $h = h_n (n = 0, 1, 2, \ldots)$ such that (10) has a Hopf bifurcation from the positive equilibrium E_+ at $h = h_n (n = 0, 1, 2, \ldots)$.*

Theorem 14 (Ma and Takeuchi 2004). *The positive equilibrium E_+ of (10) is locally asymptotically stable for $0 \leq h < h_0$, and is unstable for $h > h_0$.*

It is clear that Theorems 10 and 12 give a completed analysis on the global asymptotic properties of the solutions (10) for $b > \mu_1$ and $b \geq \mu_3(\mu_2 + \gamma)/(\mu_3 + \gamma)$. For the case of $\mu_1 = b < \mu_3(\mu_2 + \gamma)/(\mu_3 + \gamma)$, Theorem 11 shows that the disease ultimately tends to extinction. However, to give a more explicit estimation to the constant c in Theorem 11 is also an interesting problem. Theorem 14 gives a detailed analysis on the locally asymptotic properties of the positive equilibrium E_+ of (10). However, analysis of global asymptotic properties (such as permanence and global asymptotic stability) of the positive equilibrium E_+ of (10) remains to be still an important problem to be studied.

3.4 SIR epidemic models with time delays and non-constant birth rate $b(1 - \frac{\beta_1 N(t)}{1+N(t)})$

Recently, Song and Ma (2006) proposed the following delayed SIR epidemic model with density dependent birth rate:

$$\dot{S}(t) = -\beta S(t) I(t - h) - \mu_1 S(t) + b \left(1 - \beta_1 \frac{N(t)}{1 + N(t)} \right) ,$$
$$\dot{I}(t) = \beta S(t) I(t - h) - \mu_2 I(t) - \gamma I(t) , \tag{12}$$
$$\dot{R}(t) = \gamma I(t) - \mu_3 R(t) ,$$

where $S(t), I(t), R(t), N(t), \mu_1, \mu_2, \mu_3, b, \gamma$ and h have the same biological meanings as in model (10). The constant $\beta_1 (0 \leq \beta_1 < 1)$ reflects the relation between the birth rate and the population density. For $\beta_1 = 0$, the model (12) is reduced to a special case of the model (9).

It is easily to have that (12) always has a disease free equilibrium (i. e. boundary equilibrium) $E_0 = (S_0, 0, 0)$, where

$$S_0 = \frac{[b(1 - \beta_1) - \mu_1] + \sqrt{[b(1 - \beta_1) - \mu_1]^2 + 4\mu_1 b}}{2\mu_1} .$$

Furthermore, if

$$S_0 > S^* \equiv \frac{\mu_2 + \gamma}{\beta} , \tag{13}$$

then (12) also has an endemic equilibrium (i. e. interior equilibrium) $E_+ = (S^*, I^*, R^*)$, where

$$I^* = -P + \frac{\sqrt{P^2 - 4\beta S^* W Q}}{2\beta S^* W}, \qquad R^* = \frac{\gamma I^*}{\mu_3},$$

$$W = 1 + \frac{\gamma}{\mu_3} > 0,$$

$$P = [\mu_1 S^* - b(1 - \beta_1)]W + \beta S^*(1 + S^*),$$

$$Q = [\mu_1 S^* - b(1 - \beta_1)](1 + S^*) - b\beta_1 < 0.$$

3.4.1 Local asymptotic stability analysis

For local asymptotic stability of the disease free equilibrium E_0 and the endemic equilibrium E_+ of (12), we have the following results:

Theorem 15 (Song and Ma 2006). *If $S_0 < S^*$, then the disease free equilibrium E_0 of (12) is locally asymptotically stable for any time delay h. If $S_0 > S^*$, then E_0 is unstable for any time delay h.*

Theorem 16 (Song and Ma 2006). *If $S_0 > S^*$, then the endemic equilibrium E_+ of (12) is locally asymptotically stable for any time delay h.*

3.4.2 Global asymptotic stability of E_0

For global asymptotic stability of the disease free equilibrium E_0 of (12), we have the following result:

Theorem 17 (Song and Ma 2006). *If $S_0 < S^*$, the disease free equilibrium E_0 of (12) is globally asymptotically stable for any time delay h. If $S_0 = S^*$, E_0 is globally attractive for any time delay h.*

Theorem 17 shows that, while the endemic equilibrium E_+ of (12) is not feasible (i. e. $S_0 \leq S^*$), the disease free equilibrium E_0 of (12) is also globally asymptotically attractive for any time delay h. Hence, from the biological point of view, Theorems 16 and 17 suggest the following conjecture, which includes Problem 2 as a special case, may be true:

Problem 3. If $S_0 > S^*$, then the endemic equilibrium E_+ of (12) is also globally asymptotically stable for any time delay h, i. e., in biology, the inequality $S_0 > S^*$ is the threshold for an epidemic disease to occur.

3.4.3 Permanence of (12)

It seems not to be so easy to give a positive answer to Problem 3 or even to Problem 2, but we have the following result which gives a partial answer to Problem 3.

Theorem 18 (Song, Ma and Takeuchi 2005). *For any time delay h, the inequality $S_0 > S^*$ is necessary and sufficient for permanence of (12).*

Acknowledgement. The authors thank Japan Student Services Organization (JASSO) and Department of Foreign Affairs of University of Science and Technology Beijing (USTB) for their financial support.

References

1. Anderson, R. M. and R. M. May (1979), Population biology of infectious diseases: Part I, *Nature*, **280**, 361–367.
2. Beretta, E., V. Capasso and F. Rinaldi (1988), Global stability results for a generalized Lotka-Volterra system with distributed delays: Applications to predator-prey and epidemic systems, *J. Math. Biol.*, **26**, 661–668.
3. Beretta, E. and Y. Takeuchi (1995), Global stability of an *SIR* epidemic model with time delays, *J. Math. Biol.*, **33**, 250–260.
4. Beretta, E. and Y. Takeuchi (1997), Convergence results in *SIR* epidemic models with varying population size, *Nonl. Anal.*, **28**, 1909–1921.
5. Beretta, E., T. Hara, W. Ma and Y. Takeuchi (2001), Global asymptotic stability of an *SIR* epidemic model with distributed time delay, *Nonl. Anal.*, **47**, 4107–4115.
6. Butler, G., H. I. Freedman and P. Waltman (1986), Uniformly persistent systems, *Proc. Amer. Math. Soc.*, **96**, 425–430.
7. Chen, L., X. Song and Z. Lu (2002), Mathematical Models and Methods in Ecology, Sichuan Science Press, Chengdu, China.
8. Cooke, K. L. (1979), Stability analysis for a vector disease model, *Rocky Mount. J. Math.*, **9**, 31–42.
9. Cooke, K. L. and P. Van Den Driessche (1996), Analysis of an *SEIRS* epidemic model with two delays, *J. Math. Biol.*, **35**, 240–260.
10. Cooke, K. L., P. Van Den Driessche and X. Zou (1999), Interaction of maturation delay and nonlinear birth in population and epidemic models, *J. Math. Biol.*, **39**, 332–352.
11. Cushing, J. M. (1977), Integrodifferential Equations and Delays Models in Population Dynamics, Lecture Notes in Biomathematics, Vol.20, Springer-Verlag, New York.
12. Di Liddo, A. (1986), An SIR vector disease model with delay, *Math. Model*, **7**, 793–802.
13. Freedman, H. I. and S. Ruan (1995), Uniform persistence in functional differential equations, *J. Differential Equations*, **115**, 173–192.
14. Hale, J. K. (1977), Theory of Functional Differential Equations, Springer-Verlag, New York.
15. Hale, J. K. and P. Waltman (1989), Persistence in infinite- dimensional systems, *SIAM J. Math. Anal.*, **20**, 388–395.
16. Hethcote, H. W. (1976), Qualitative analyses of communicable disease models, *Math. Biosci.*, **7**, 335–356.
17. Hethcote, H. W. and P. Van Den Driessche (2000), Two *SIS* epidemiological models with delays, *J. Math. Biol.*, **40**, 2–26.

18. Kermack, W. O. and A. G. McKendrick (1927), A contribution to the mathematical theory of epidemics, *Proc. Roy. Soc.*, **A115**, 700–721.

19. Kuang, Y. (1993), Delay Differential Equations with Applications in Population Dynamics, Academic Press, New York.

20. Liu, W., S. A. Levin and Y. Iwasa (1986), Influence of nonlinear incidence rate upon the behavior of $SIRS$ epidemiological models, *J. Math. Biol.*, **23**, 187–204.

21. Ma, W. and Y. Takeuchi (2004), Asymptotic properties of a delayed SIR epidemic model with density dependent birth rate, *Discrete Cont. Dyn. Sys.*, **4**, 671–678.

22. Ma, W., Y. Takeuchi, T. Hara and E. Beretta (2002), Permanence of an SIR epidemic model with distributed time delays, *Tohoku J. Math.*, **54**, 581–591.

23. Ma, W., M. Song and Y. Takeuchi (2004), Global stability of an SIR epidemic model with time delay, *Appl. Math. Letters*, **17**, 1141–1145.

24. Ma, Z., Y. Zhou, W. Wang and Z. Jin (2004), Mathematical Models in Epidemiology Dynamics, Science Press, Beijing.

25. McDonald, N. (1978), Time Lags in Biological Models, Lecture Notes in Biomathematics, Vol.27, Springer-Verlag, New York.

26. Mena-Lorca, J. and H. W. Hethcote (1992), Dynamic models of infectious diseases as regulators of population sizes, *J. Math. Biol.*, **30**, 693–716.

27. Ruan, S. (2007), Spatial-temporal dynamics in nonlocal epidemiological models, Mathematics for Life Science and Medicine, Eds. Y. Takeuchi et al, Springer-Verlag, Berlin Heidelberg, 107–122.

28. Smith, H. L. (1983), Subharmonic bifurcation in an SIR epidemic model, *J. Math. Biol.*, **17**, 163–177.

29. Song, M. and W. Ma (2006), Asymptotic properties of a revised SIR epidemic model with density dependent birth rate and time delay, *Dyn. Conti. Discrete Impul. Sys.*, **13**, 199–208.

30. Song, W. Ma and Y. Takeuchi (2005), Permanence of a delayed SIR epidemic model with density dependent birth rate, *J. Compt. Appl. Math.*, (to appear).

31. Takeuchi, Y. and W. Ma (1999), Stability analysis on a delayed SIR epidemic model with density dependent birth process, *Dyn. Conti. Discrete Impul. Sys.*, **5**, 171–184.

32. Takeuchi, Y., W. Ma and E. Beretta (2000), Global asymptotic properties of a delayed SIR epidemic model with finite incubation time, *Nonl. Anal.*, **42**, 931–947.

33. Thieme, H. R. (2003), Mathematics in Population Biology, Princeton University Press.

34. Wang, W. (2002), Global behavior of an SEIRS epidemic model with time delay, *Appl. Math. Letters*, **15**, 423–428.

4

Epidemic Models with Population Dispersal

Wendi Wang

Summary. The purpose of this chapter is to outline recent advances of mathematical models in the studies of epidemic diseases in heterogeneous geography. Section 4.1 presents a model with the immigration of infectives from outside the population. Section 4.2 introduces multi patches into epidemic models but assume that the population size in each patch is constant in time. One objective is to discuss conditions under which patches become synchronised. The second objective is to consider the invasion of malaria into a population distributed in distinct patches. In Sect. 4.3, a demographic structure is incorporated into the epidemic models with multi patches. The basic reproduction number of the model is established. Section 4.4 changes the mass action incidence to a standard incidence. Section 4.5 further considers the residence of individuals, which make mathematical models more accurate.

4.1 Introduction

Many mathematical models have been proposed to understand the mechanism of disease transmission. One way for this purpose is to improve classical models to see the effect of various biological factors on propagation of a disease. First, we can consider nonlinear incidences caused by behavior changes or nonlinear treatment rates (see, for example, Liu et al.(1986, 1987), Ruan and Wang (2003), Wang and Ruan (2004b)). Secondly, we can introduce time delays from the latent period, infection period and recovery period (Cooke and van den Driessche (1996), Feng and Thieme (2000a, 2002b), Beretta et al. (2000, 2002), Wang and Ma (2002a, 2002b)), or consider an age effect (Gyllenberg and Webb (1990), Inaba (1990)). We can also consider disease transmissions in multiple populations. For example, Xiao and Chen (2001) studied disease transmissions in a predator-prey system, Han et al. (2003) studied disease transmissions in a system of competing species. Most of these papers assume homogeneous space distribution both for populations and disease transmissions.

Space structure plays an important role in the spread of a disease. Some epidemic diseases have occurred in some regions frequently and were transmitted to other regions due to population dispersal. For example, SARS was first reported in Guangdong Province of China in November of 2002. The emerging disease spread very quickly to some other regions in Mainland China as well as Hong Kong, Singapore, Vietnam, Canada, etc. In March of 2003, the World Health Organization, for the first time in its history, issued a globally warning about the disease. In late June of 2003, the disease was under control globally, but it had spread to 32 countries and regions causing about 800 deaths and more than 8000 infections (see, for example, Wang and Ruan (2004a)). Hence, it is important to use mathematical models to understand the effect of population dispersal on the spread of a disease. Basically, there are two ways for this purpose. First, we can introduce space variables and use reaction diffusion equations (see Murray (1989) and Britton (2003) and the references cited therein). One of the major limitations of diffusion models is the assumption of the movements of individuals among direct neighborhoods. In nature many organisms can move or can be transferred over large distances. For example, birds can fly to remote habitats, and human populations can move into other countries in a short time due to modern transportation tools. Further, many populations live in the form of communities, for example, human population live in cities. Thus, it is reasonable to adopt, as an alternative, patch models, in epidemiology. Here, one patch may represent a city or or a biological habitat.

4.2 Epidemic models with immigration of infectives

Communicable diseases may be introduced into a population by the arrival of infectives from outside the population. For example, travellers may return home from a foreign trip with an infection acquired abroad, or individuals who are HIV positive may enter into a prison. For these reasons, Brauer and van den Driessche (2001) proposed an epidemic model with immigration of infectives. They consider an SIS model. Let $S(t)$ be the number of members of a population who are susceptible to an infection at time t, $I(t)$ the number of members who are infective at time t. It is assumed that there is a constant flow A of new members into the population in unit time, of which a fraction p $(0 \le p \le 1)$ is infective. The model is

$$\begin{aligned}
S' &= (1 - p)A - \beta SI - dS + \gamma I \,, \\
I' &= pA + \beta SI - (d + \gamma + \alpha)I \,,
\end{aligned} \tag{1}$$

where β is the disease transmission coefficient, d is a natural death rate, α is the disease-induced death rate and γ is the recovery rate.

Advantages of (1) are that only one patch is involved and infectives from outside are also included. These make the mathematical analysis much easier.

Writing $N = S + I$, we can transform (1) into

$$I' = pA + \beta I(N - I) - (d + \gamma + \alpha)I \,,$$
$$N' = A - dN - \alpha I \,. \tag{2}$$

If $\beta = 0$, which means that only infectives are those who have entered the population from outside, it is easy to see that every solution approaches the endemic equilibrium (I_0, N_0) where

$$I_0 = \frac{pA}{d + \gamma + \alpha} \,, \quad N_0 = \frac{A}{d}\,\frac{d + \gamma + \alpha(1 - p)}{d + \gamma + \alpha} \,.$$

If $\beta > 0$, for $p > 0$, (2) has a unique endemic equilibrium (I^*, N^*) where

$$I^* = \frac{\sigma + \sqrt{\sigma^2 + 4\beta A\, dp(d + \alpha)}}{2\beta(d + \alpha)} \,,$$
$$N^* = (A - \alpha I^*)/d \,,$$
$$\sigma = \beta A - d(d + \gamma + \alpha) \,.$$

By the Bendixson-Dulac criterion, the endemic equilibrium is globally stable. Then it is easy to see that the number of infectives can be reduced by reducing p or A (i. e., the number of infectives entering the population) or by increasing γ (the recovery rate constant).

4.3 Constant population sizes in each patch

Multi-patch models with epidemic diseases were studied by Lloyd and May (1996, 2004) and Rodríguez and Torres-Sorando (2001). They consider n patches and assume that the population size of each patch remains constant even though individuals move among patches. This implies that the number of births exactly balances the number of deaths in every patch and that individuals do not move permanently from one patch to another.

4.3.1 An $SEIS$ model

In the simplest case of one patch, an $SEIR$ model is:

$$\frac{dS}{dt} = \mu N - \mu S - \lambda S \,,$$
$$\frac{dE}{dt} = \lambda S - (\mu + \sigma)E \,,$$
$$\frac{dI}{dt} = \sigma E - (\mu + \gamma)I \,, \tag{3}$$
$$\frac{dR}{dt} = \gamma I - \mu R \,,$$
$$\lambda = \beta I \,,$$

where S, E, I and R represent the numbers of susceptible, exposed (but not yet infectious), infectious, and recovered individuals respectively, μ is the birth rate and death rate, $1/\sigma$ is the average latent period of the disease, $1/\gamma$ is the average infectious period, λ is the infection force.

For n patches, it is assumed in paper (Lloyd and May 1986) that the dynamics of disease transmission in patch i is governed by

$$
\begin{aligned}
\frac{dS_i}{dt} &= \mu_i N_i - \mu_i S_i - \lambda_i S_i \,, \\
\frac{dE_i}{dt} &= \lambda_i S_i - (\mu_i + \sigma_i)E_i \,, \\
\frac{dI_i}{dt} &= \sigma_i E_i - (\mu_i + \gamma_i)I_i \,, \\
\lambda_i &= \sum_{i=1}^{n} \beta_{ij} I_j \,,
\end{aligned}
\tag{4}
$$

where S_i, E_i, I_i and R_i represent the numbers of susceptible, exposed (but not yet infectious), infectious, and recovered individuals in patch i, respectively; the parameters with subscript i admit similar meanings as those without subscript i; the equation of R_i is dropped because R_i is independent of other variables, Further, N_i is a positive constant. Let Φ be a matrix whose entries equal β_{ij}. Φ describes the disease transmission between and within patches. The formulation of the model assumes that there is an epidemiological cross-coupling between patches, but that individuals do not migrate between patches. This might be thought of as arising from a situation in which individuals make short lived visits from their home patch to other patches.

Consider a matrix $T = (T_{ij})$ where

$$
T_{ij} = \frac{\beta_{ij} N_i \sigma}{(\mu + \sigma)(\mu + \gamma))} \,.
$$

Set

$$
s(T) = \max\{\mathrm{Re}\lambda : \lambda \text{ is an eigenvalue of } T\} \,.
$$

By Lajmanovich & Yorke (1976), if $s(T) < 1$, (4) has a stable disease free equilibrium; if $s(T) > 1$, (4) has a unique endemic equilibrium (S^*, E^*, I^*).

In order to find how the system approaches the endemic equilibrium, we set $\lambda_i^* = \sum_{i=1}^{n} \beta_{ij} I_j^*$. Let $A(\Lambda)$ be a matrix whose entry $A_{ij}(\Lambda)$ is defined by

$$
A_{ij} = (\Lambda + \mu + \lambda^*)\delta_{ij} - \frac{\sigma(\Lambda + \mu)}{(\Lambda + \mu + \gamma)(\Lambda + \mu + \sigma)} S_i^* \beta_{ij} \,,
$$

where $\delta_{ii} = 1$ and $\delta_{ij} = 0$ if $i \neq j$. Then the characteristic equation of (4) at the endemic equilibrium is $\det A = 0$. Its roots determine the stability of the endemic equilibrium. In order to examine the stability analytically, we need to have analytical expressions for S_i^*, E_i^*, I_i^*. For this purpose, paper Lloyd and May (1986) considers a simplified case where each patch is of the same population size ($N_i = N$). Further, the contact rate is the same within each patch, and another (usually different and smaller) rate between each pair of distinct patches, which are given by

$$\beta_{ij} = \begin{cases} \beta, & \text{if} i = j , \\ \epsilon\beta, & \text{otherwise} \end{cases}$$

with $0 \le \epsilon \le 1$. Then, for a special case where (4) is reduced to an SIR model (letting $\sigma \to \infty$), the Φ matrix has eigenvalues given by

$$\Gamma = \beta(1 - \epsilon)$$

repeated $n - 1$ times, and

$$\Gamma = \beta(n\epsilon + 1 - \epsilon) .$$

A characteristic root Λ satisfies the following equation

$$(\Lambda + \mu R_0)\frac{\Lambda + \mu + \gamma}{(\Lambda + \mu)N}R_0 = \Gamma , \tag{5}$$

where

$$R_0 = \frac{N}{\mu + \gamma}\sum_{j=1}^{n}\beta_{ij} .$$

Since Λ can be solved from (5), paper (Lloyd and May 1986) proved that the endemic equilibrium is stable and the patches often oscillate in phase for all but the weakest between patch coupling.

4.3.2 A malaria model

Now, we suppose that the habitat is partitioned into k patches. N_i and $X_i(t)$ are the numbers of humans and infected humans, respectively, at time t in patch $i(i = 1, \ldots, k)$, M_i and $Y_i(t)$ represent the number of mosquitoes and infected mosquitoes, respectively, at time t in patch $i(i = 1, \ldots, k)$. Let us assume that humans can travel between patches but mosquitoes can not. Further, we assume that a fraction v_{ij} of the time devoted by humans to reside in patch i per unit time, is devoted to visit patch $j(j \neq i, j = 1, \ldots, k)$. After the visit, these humans return to their home patches. Under the above assumptions, paper (Rodríguez and Torres-Sorando 2001) proposed the following model:

$$\frac{\mathrm{d}X_i}{\mathrm{d}t} = b(N_i - X_i(t))Y_i(t) - gX_i(t) + b(N_i - X_i(t)) \sum_{j \neq i} v_{ij}Y_j(t) \,,$$

$$\frac{\mathrm{d}Y_i}{\mathrm{d}t} = b(M_i - Y_i(t))X_i(t) - mY_i(t) + b(M_i - Y_i(t)) \sum_{j \neq i} v_{ji}X_j(t) \,, \qquad (6)$$

$$i = 1, \ldots, k \,,$$

where b is the disease transmission coefficient due to contacts from susceptible humans and infectious mosquitoes, or susceptible mosquitoes and infectious humans, g is the recovery rate of infected human individuals, m is the recovery rate of infected mosquitoes.

Tow patterns of spatial array were considered in Rodríguez and Torres-Sorando (2001). The first is a unidimensional one, in which the k patches are arranged as cells in a row. This array can simulate the spatial distribution of patches along a coast. The second array is a bidimensional one, in which the k patches are arranged in a rectangle whose four sides are of the smae length or number of cells. This bidimensional array can simulate a general spatial distribution on a land surface.

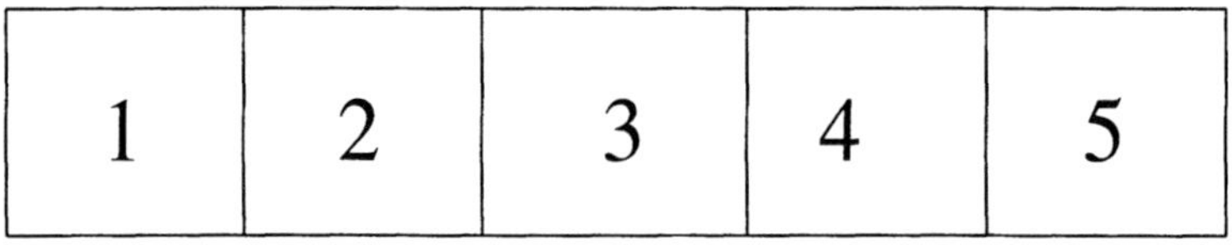

1	2	3	4	5

Fig. 4.1. Unidimensional array with $k = 5$

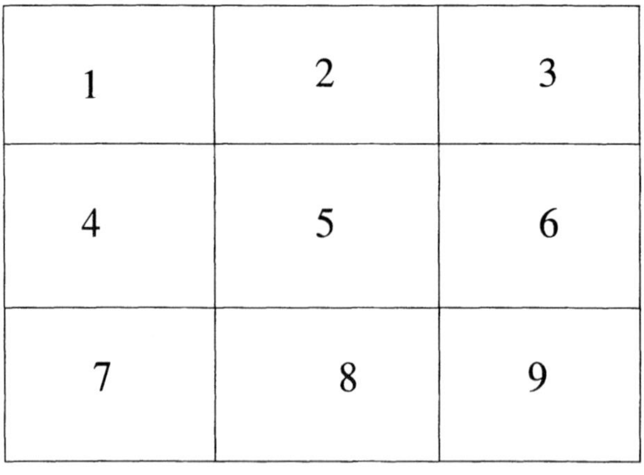

1	2	3
4	5	6
7	8	9

Fig. 4.2. Bidimensional array with $k = 9$

If $v_{ij} = v$ for visitation $(i \neq j, i, j = 1, \ldots, k)$, this means that the magnitude of contact between patches is constant, no matter how far apart they are. In order to incorporate distance effects on contacts between patches, we assume that v is the probability of visiting a patch one distance unit away per unit time, and $v_{ij} = v^{e_{ij}}$ with e_{ij} being the number of distance units between patch i and patch $j (i \neq j, i, j = 1, \ldots, k)$.

Let E be a matrix whose entries are e_{ij}. For the unidimensional case,

$$e_{ij} = \sqrt{(i-j)^2} \,.$$

For example, for $k = 5$ we have

$$E = \begin{bmatrix} 0 & 1 & 2 & 3 & 4 \\ 1 & 0 & 1 & 2 & 3 \\ 2 & 1 & 0 & 1 & 2 \\ 3 & 2 & 1 & 0 & 1 \\ 4 & 3 & 2 & 1 & 0 \end{bmatrix}.$$

For the bidimensional case the elements of the matrix E have to be obtained for each k. For example for $k = 9$, that is, an area of a rectangle is partitioned into 9 cells, $e_{12} = 1, e_{13} = 2, e_{15} = \sqrt{1^2 + 1^2} = \sqrt{2}$, and we can write the following:

$$E = \begin{bmatrix} E_1 & E_2 & E_3 \\ E_2 & E_1 & E_2 \\ E_3 & E_2 & E_1 \end{bmatrix},$$

where

$$E_1 = \begin{bmatrix} 0 & 1 & 2 \\ 1 & 0 & 1 \\ 2 & 1 & 0 \end{bmatrix}, \quad \begin{bmatrix} 1 & \sqrt{2} & \sqrt{5} \\ \sqrt{2} & 1 & \sqrt{2} \\ \sqrt{5} & \sqrt{2} & 1 \end{bmatrix},$$

and

$$E_3 = \begin{bmatrix} 2 & \sqrt{5} & \sqrt{8} \\ \sqrt{5} & 2 & \sqrt{5} \\ \sqrt{8} & \sqrt{5} & 2 \end{bmatrix}.$$

Let N be the total number of humans and M be the total number of mosquitoes. We now assume that the total number of individuals (humans and mosquitoes are initially evenly distributed. Since we also assume that the pattern of movement between patches does not produce any net change in the number of each patch, the number of humans in each patch will be

N/k, and that of mosquitoes M/k. Then (6) simplifies to

$$\frac{\mathrm{d}X_i}{\mathrm{d}t} = b\left(\frac{N}{k} - X_i(t)\right)Y_i(t) - gX_i(t) + b\left(\frac{N}{k} - X_i(t)\right)\sum_{j\neq i} v_{ij}Y_j(t)\,,$$

$$\frac{\mathrm{d}Y_i}{\mathrm{d}t} = b\left(\frac{M}{k} - Y_i(t)\right)X_i(t) - mY_i(t) + b\left(\frac{M}{k} - Y_i(t)\right)\sum_{j\neq i} v_{ji}X_j(t)\,, \qquad (7)$$

$$i = 1,\ldots,k\,.$$

Paper (Rodríguez and Torres-Sorando 2001) obtained conditions that the malaria can invade the human population in a sense that the disease free equilibrium of (7) is unstable. If there is no effect of distance, the condition is

$$b\sqrt{\frac{NM}{gm}} > \frac{k}{1+(k-1)v}\,. \qquad (8)$$

If there is the effect of distance given by $v_{ij} = v^{e_{ij}}$ with e_{ij} being the number of distance units between patch i and patch j, for $k \gg 1$ in the one-dimensional case, the condition for invasion of the disease is

$$b\sqrt{\frac{NM}{gm}} > \frac{k(1-v)}{1+(k-1)v}\,. \qquad (9)$$

Equations (8) and (9) imply that the establishment of the disease is more restricted as the environment is more partitioned.

4.4 Multi-patches with demographic structure

In this section we relax the assumption that population size in each patch is constant. Thus, the timescale of an epidemic disease is allowed to be longer and individuals can move from one patch to other patches. That is, there are immigrations and emigrations.

4.4.1 Formulation of patch models

We consider n patches. First, we have to choose a demographic structure in each patch. For simplicity, We assume that the population dynamics in patch i is described by

$$N_i' = B(N_i)N_i - \mu_i N_i\,,$$

where N_i is the number (or density) of a population in patch i, $B_i(N_i)$ is the per capita birth rate of the population in patch i, μ_i is the death rate of individuals in patch i. Here, we have assumed that the regulation effect of population density only occurs in the birth process, but not in the death process. This type of demographic structure with variable population size was proposed by Cooke et al. (1999). We assume that $B_i(N_i)$ satisfy the following basic assumptions for $N_i \in (0, \infty)$:

(A1) $B_i(N_i) > 0, i = 1, 2, \ldots, n$;
(A2) $B_i'(N_i)$ is continuous and $B_i'(N_i) < 0, i = 1, 2, \ldots, n$;
(A3) $\mu_i > B_i(\infty), i = 1, 2, \ldots, n$.

(A1) means that the per capita birth rate is positive, (A2) indicates that it is a decreasing function of population density, (A3) implies that the net growth rate of the population is negative when population density is large, which prevents an unbounded population size. According to Cooke et al. (1999), we can adopt at least the following three types of birth functions B_i:

(B1) $B_i(N_i) = b_i e^{-a_i N_i}$ with $a_i > 0, b_i > 0$;
(B2) $B_i(N_i) = \frac{p_i}{q_i + N_i^m}$ with $p_i, q_i, m > 0$;
(B3) $B_i(N_i) = \frac{A_i}{N_i} + c_i$ with $A_i > 0, c_i > 0$.

(B1) means that the birth process obeys Ricker's law. (B2) indicates that the birth process is of Beverton-Holt type. In the type of (B3), the population dynamics is given by

$$N_i' = A_i - (\mu_i - c_i)N_i \, .$$

This type of demographic structure has been adopted by many papers in the literature.

   If we consider an *SIS* type of disease transmission, the population is divided into two classes: susceptible individuals and infectious individuals. Susceptible individuals become infective after contact with infective individuals. Infective individuals return to the susceptible class when they are recovered. Gonorrhea and other sexually transmitted diseases or bacterial infections exhibit this phenomenon. Let S_i be the number (or density) of susceptible individuals in patch i, I_i the number (or density) of infectious individuals in patch i, $N_i = S_i + I_i$ the number (or density) of the population in patch i, $B_i(N_i)$ the birth rate of the population in the patch i, μ_i the death rate of the population in the patch i, and γ_i the recovery rate of infective individuals in the patch i. Suppose that the infection force (the probability that a susceptible is infected in unit time) in patch i is given by $\phi_i(I_i, N_i)$. Several infection forces are frequently used in literature. If $\phi_i(I_i, N_i) = \beta_i I_i$, we have a mass action incidence $\beta_i I_i S_i$. If $\phi_i(I_i, N_i) = \beta_i I_i / N_i$, we adopt a standard incidence $\beta_i S_i I_i / N_i$. The infection force $\phi_i(I_i, N_i) = \beta_i c(N_i) I_i / N_i$ was used by Diekmann and Heesterbeek (2000) where $c(N_i)$ is the encounter number of one individual with other members per unit time. The infection force $\phi_i(I_i, N_i) = \beta_i(I_i)$ was considered by Liu et al. (1986), Liu et al. (1987), van den Driessche and Watmough (2000), and Ruan and Wang (2003). We assume that the infection force ϕ_i satisfies:

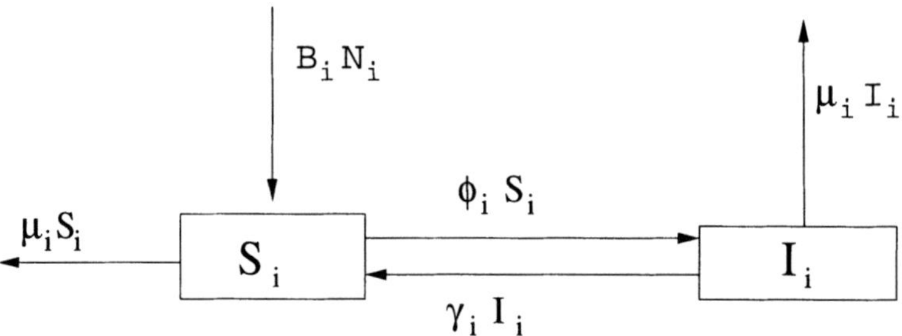

Fig. 4.3. Scheme of population demography and disease transmission in patch i

(C1) $\phi_i(0, N_i) = 0$, ϕ_i is continuously differentiable with respect to I_i and N_i for $I_i \geq 0, 0 < N_i < \infty$.

If there is no population dispersal among patches, i. e., the patches are isolated, we suppose that the population dynamics in i-th patch is governed by

$$\begin{cases} S_i' = B_i(N_i)N_i - \mu_i S_i - \phi_i(I_i, N_i)S_i + \gamma_i I_i \, , \\ I_i' = \phi_i(I_i, N_i)S_i - (\mu_i + \gamma_i)I_i \, . \end{cases} \qquad (10)$$

We now consider population dispersal among the patches. Let $-a_{ii}$ represent the emigration rate of susceptible individuals in the i-th patch and $-b_{ii}$ represent the emigration rate of infective individuals in the i-th patch, where $a_{ii}, b_{ii}, 1 \leq i \leq n$, are non-positive constants. Further, let $a_{ij}, j \neq i$, represent the immigration rate of susceptible individuals from the j-th patch to the i-th patch, and $b_{ij}, j \neq i$, the immigration rate of infective individuals from the j-th patch to the i-th patch. For simplicity, we neglect death rates and birth rates of individuals during their dispersal process, and any quarantine or culling for infectious individuals in the paths of migration. Then, we have

$$\sum_{j=1}^{n} a_{ji} = 0 \, , \quad \sum_{j=1}^{n} b_{ji} = 0 \, , \quad \forall 1 \leq i \leq n \, . \qquad (11)$$

Hence, when the patches are connected, we have the following epidemic model with population dispersal (see Fig. 4.4):

$$\begin{cases} S_i' = B_i(N_i)N_i - \mu_i S_i - \phi_i(I_i, N_i)S_i + \gamma_i I_i + \sum_{j=1}^{n} a_{ij}S_j \, , \\ I_i' = \phi_i(I_i, N_i)S_i - (\mu_i + \gamma_i)I_i + \sum_{j=1}^{n} b_{ij}I_j \, , \\ 1 \leq i \leq n \, . \end{cases} \qquad (12)$$

We assume that the n patches cannot be separated into two groups such that there is no immigration of susceptible and infective individuals from the first group to the second group. Mathematically, this means that the two $n \times n$ matrices (a_{ij}) and (b_{ij}) are irreducible (see, e. g., appendix A (Smith and

Waltman 1995)). Note that system (12) indicates that the population can have different demographic structures and different infection forces among different patches. This means that we have included the space distributions of demographic factors and epidemic factors into the epidemic model. Further, a_{ij} may be different from b_{ij}. If $b_{ij} < a_{ij}$, this means that we perform control measures on the movement of infectives. Especially, $b_{ij} = 0$ means that infectious individuals can not move from the j-th patch to the i-th patch due to strict screening of infected individuals at the borders between patch i and patch j. On the other hand, $b_{ij} > a_{ij}$ may mean terrorism activities.

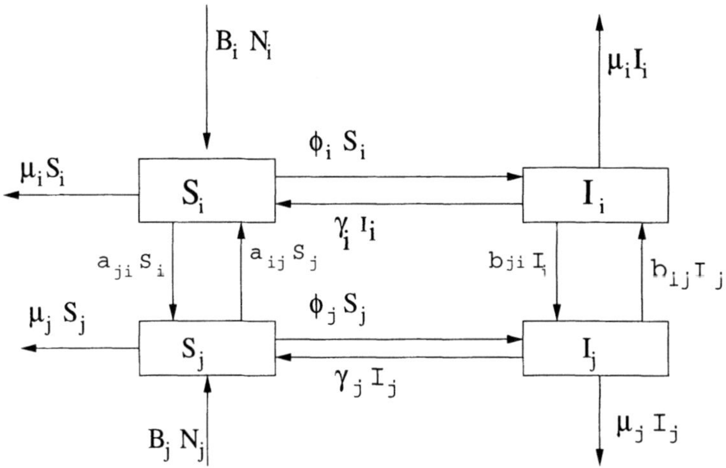

Fig. 4.4. Scheme of population birth, death, movement and disease transmission

4.4.2 Basic reproduction number

Our main concern now is the basic reproduction number of (12). If we introduce a typical infective into a completely susceptible population, the number of new infectives produced by this single infective during its infection period is called as a basic reproduction number. If the number of patches is one and the carrying capacity of the population is K, then the basic reproduction number $R_0 = \phi(0, K)/(\mu + \gamma)$, in which $\phi(0, K)$ is per capita infection rate(at disease free equilibrium) and $1/(\mu+\gamma)$ is the infection period. In order to obtain the basic reproduction number for multiple patches, first, we find a disease free equilibrium $(S^*, 0)$ of (12). S^* is a positive equilibrium of the following system

$$S_i' = B_i(S_i)S_i - \mu_i S_i + \sum_{j=1}^{n} a_{ij}S_j, \quad i = 1, \ldots, n. \tag{13}$$

In order to obtain the existence and uniqueness of S^*, set

$$M(0) = \begin{bmatrix} B_1(0) - \mu_1 + a_{11} & a_{12} & \cdots & a_{1n} \\ a_{21} & B_2(0) - \mu_2 + a_{22} & \cdots & a_{2n} \\ \cdots & \cdots & \cdots & \cdots \\ a_{n1} & a_{n2} & \cdots & B_n(0) - \mu_n + a_{nn} \end{bmatrix}.$$

Assume

(A4) $s(M(0)) > 0$, where

$$s(M(0)) = \max\{Re\lambda \colon \lambda \text{ is an eigenvalue of } M(0)\}.$$

Note that (13) is cooperative, i.e., the flow of (13) is monotonic with respect to initial positions. Since (A1)–(A4) imply that positive solutions of (13) are bounded, the origin repels positive solutions of (13) and the right-hand side is sublinear on R_+^n, it follows from the theory of monotonic flow (for example, see Corollary 3.2 (Zhao and Jing 1996)) that (13) has a unique positive equiulibrium $S^* = (S_1^*, S_2^*, \ldots, S_n^*)$ which is globally asymptotically stable for $S \in R_+^n \setminus \{0\}$. Thus, $E_0 = (S_1^*, S_2^*, \ldots, S_n^*, 0, \ldots, 0)$ is a unique disease free equilibrium of (12).

Next, we define a matrix

$$F := \begin{bmatrix} \xi_1 S_1^* & 0 & \cdots & 0 \\ 0 & \xi_2 S_2^* & \cdots & 0 \\ \cdots & \cdots & \cdots & \cdots \\ 0 & 0 & \cdots & \xi_n S_n^* \end{bmatrix}$$

where $\xi_i = \frac{\partial}{\partial I_i}\phi_i(I_i, S_i + I_i)|_{S_i=S_i^*, I_i=0}$. Here, F represents the infection rate matrix for the patches. In order to calculate the distribution of an infective staying in every patch, we define a matrix V by

$$V = - \begin{bmatrix} -\mu_1 - \gamma_1 + b_{11} & b_{12} & \vdots & b_{1n} \\ b_{21} & -\mu_2 - \gamma_2 + b_{22} & \vdots & b_{2n} \\ \cdots & \cdots & \vdots & \cdots \\ b_{n1} & b_{n2} & \vdots & -\mu_n - \gamma_n + b_{nn} \end{bmatrix}.$$

Here, $-V$ represents the transfer rates of infected individuals among patches due to death, recovery and movement. According to van den Driessche and Watmough (2002), V^{-1} gives the distribution of an infected individual staying in every patch, that is, if we introduce an infective into patch k, then the (j, k) entry of V^{-1} is the average length of time this infective stays in patch j. As a consequence, FV^{-1} is the next generation matrix. By Diekmann et al.(1990),

$$R_0 := \rho(FV^{-1}),$$

is the basic reproduction number for (12), where ρ represents the spectral radius of the matrix.

By modifying the proof of Wang and Zhao (2004) and using the theory of monotonic flow (Smith 1995) and the theory of persistence (Freedman and Waltman 1984; Thieme 1993), we can obtain

Theorem 1. *Let (A1)–(A4), (C1) hold and $R_0 < 1$. Then the disease free equilibrium is locally asymptotically stable.*

Theorem 2. *Let (A1)–(A4), (C1) hold and $R_0 > 1$. Then there is a positive constant ϵ such that every solution $(S(t), I(t))$ of (12) with $(S(0), I(0)) \in R_+^n \times int(R_+^n)$ satisfies*

$$\liminf_{t \to \infty} I_i(t) \geq \epsilon, \quad i = 1, 2, \ldots, n \,.$$

4.4.3 A model of two patches with mass action incidence

In order to illustrate the effect of population dispersal on the disease spread, we consider a special case where the patch number is 2, the birth functions B_i take the form:

$$B_i(N_i) = \frac{r_i}{N_i} + c_i \,, \quad c_i < \mu_i \,, \quad 1 \leq i \leq 2 \,,$$

and mass action incidences are adopted (Wang and Zhao 2004). Then (12) becomes

$$\begin{aligned}
S_1' &= r_1 + c_1 I_1 - (\mu_1 - c_1 + a_{11})S_1 - \beta_1 S_1 I_1 \\
&\quad + \gamma_1 I_1 + a_{22} S_2 \\
I_1' &= \beta_1 S_1 I_1 - (\mu_1 + \gamma_1 + b_{11})I_1 + b_{22} I_2 \\
S_2' &= r_2 + c_2 I_2 - (\mu_2 - c_2 + a_{22})S_2 - \beta_2 S_2 I_2 \\
&\quad + \gamma_2 I_2 + a_{11} S_1 \\
I_2' &= \beta_2 S_2 I_2 - (\mu_2 + \gamma_2 + b_{22})I_2 + b_{11} I_1 \,.
\end{aligned} \tag{14}$$

Advantage for this system is that the disease free equilibrium $E_0 = (S_1^*, S_2^*, 0, 0)$ is given explicitly by

$$S_1^* = \frac{\mu_2\, r_1 - c_2\, r_1 + a_{22}\, r_1 + a_{22}\, r_2}{\mu_1\, \mu_2 - \mu_1\, c_2 + \mu_1\, a_{22} - c_1\, \mu_2 + c_1\, c_2 - c_1\, a_{22} + a_{11}\, \mu_2 - a_{11}\, c_2} \,,$$

$$S_2^* = \frac{a_{11}\, r_1 + \mu_1\, r_2 - c_1\, r_2 + a_{11}\, r_2}{\mu_1\, \mu_2 - \mu_1\, c_2 + \mu_1\, a_{22} - c_1\, \mu_2 + c_1\, c_2 - c_1\, a_{22} + a_{11}\, \mu_2 - a_{11}\, c_2} \,.$$

Therefore, the basic reproduction number can be given explicitly by $R_0 = \rho(FV^{-1})$ where

$$F = \begin{bmatrix} \beta_1 S_1^* & 0 \\ 0 & \beta_2 S_2^* \end{bmatrix}, \quad V = -\begin{bmatrix} -\mu_1 - \gamma_1 - b_{11} & b_{22} \\ b_{11} & -\mu_2 - \gamma_2 - b_{22} \end{bmatrix}$$

In the absence of population dispersal between two patches, that is, $a_{11} = a_{22} = b_{11} = b_{22} = 0$, (14) becomes

$$S_1' = r_1 + c_1 I_1 - (\mu_1 - c_1)S_1 - \beta_1 S_1 I_1 + \gamma_1 I_1$$
$$I_1' = \beta_1 S_1 I_1 - (\mu_1 + \gamma_1)I_1 \tag{15}$$

and

$$S_2' = r_2 + c_2 I_2 - (\mu_2 - c_2)S_2 - \beta_2 S_2 I_2 + \gamma_2 I_2$$
$$I_2' = \beta_2 S_2 I_2 - (\mu_2 + \gamma_2)I_2 \, . \tag{16}$$

Set

$$R_{01} := \frac{\beta_1 r_1}{(\mu_1 - c_1)(\mu_1 + \gamma_1)} \, , \tag{17}$$

$$R_{02} := \frac{\beta_2 r_2}{(\mu_2 - c_2)(\mu_2 + \gamma_2)} \, . \tag{18}$$

R_{01} is a basic reproduction number of the disease in the first patch. Since (15) is two dimensional, where the Bendixson Theorem applies, it is easy to verify that the disease will disappear in the first patch if $R_{01} < 1$ and there is an endemic equilibrium in (15) which is globally asymptotically stable if $R_{01} > 1$. Similarly, R_{02} is a basic reproduction number of the disease in the second patch. the disease will disappear in the second patch if $R_{02} < 1$ and there is an endemic equilibrium in (16) which is globally asymptotically stable if $R_{02} > 1$.

Now, we suppose that the patches are connected. Then the population dispersal may facilitate the spread of the disease or reduce the risk of the disease spread. This can be seen from the following theorems and examples.

Example 1. Suppose $r_1 = r_2 = r, c_1 = c_2 = c, \mu_1 = \mu_2 = \mu, \gamma_1 = \gamma_2 = \gamma, a_{11} = a_{22} = b_{11} = b_{22} = \theta$ in (14). This means that we adopt the same demographic structure, same recovery rate, and same migration rate in two patches. We vary θ, β_1, β_2 to see the effect of the contact rates and the dispersal rate on the disease spread. Then

$$S_1^* = \frac{r}{\mu - c} \, , \qquad S_2^* = \frac{r}{\mu - c} \, ,$$
$$R_{01} = \frac{S_1^* \beta_1}{\mu + \gamma} \, , \qquad R_{02} = \frac{S_2^* \beta_2}{\mu + \gamma} \, .$$

Assume that the disease spreads in each isolated patch, i. e., $R_{0i} > 1, i = 1, 2$. Notice that the characteristic equation of matrix FV^{-1} in this case is

$$\lambda^2 - q_1 \lambda + q_2 = 0 \, , \tag{19}$$

where

$$q_1 = \frac{S_1^* (\mu + \gamma + \theta)(\beta_1 + \beta_2)}{(\mu + \gamma)(\mu + \gamma + 2\theta)} \, , \qquad q_2 = \frac{\beta_1 S_1^* S_2^* \beta_2}{(\mu + \gamma)(\mu + \gamma + 2\theta)} \, .$$

For (19), necessary and sufficient conditions for $|\lambda_i| < 1, i = 1, 2$, are the Jury conditions (Britton 2003):

$$|q_1| < q_2 + 1 , \quad q_2 < 1 . \tag{20}$$

If

$$\theta < \frac{1}{2} \frac{\beta_1 \, S^2 \, \beta_2 - \mu^2 - 2\mu\gamma - \gamma^2}{\mu + \gamma} ,$$

we have $q_2 > 1$. It follows from the Jury conditions that $R_0 > 1$. If

$$\theta \geq \frac{1}{2} \frac{\beta_1 \, S^2 \beta_2 - \mu^2 - 2\mu\gamma - \gamma^2}{\mu + \gamma} ,$$

it is not hard to verify that $q_2 + 1 < q_1$. Again, the Jury conditions imply that $R_0 > 1$. It follows that the disease also spreads in two patches when population dispersal occurs.

Now, we suppose that the disease dies out in each isolated patch, i.e., $R_{0i} < 1, i = 1, 2$. In this case, we have $q_2 < 1$. Then arguing as above, we can verify that $q_2 + 1 > q_1$. It follows from the Jury conditions that $R_0 < 1$. Thus, the disease dies also out in the two patches when population dispersal occurs. These two cases are expected.

Now, we fix $r = c = \gamma = 1, \mu = 2, \beta_2 = 1, \beta_1 = 6$. Then $R_{01} = 2, R_{02} < 1/3$. The characteristic equation becomes

$$\lambda^2 - \frac{7}{3} \frac{3+\theta}{3 + 2\theta} \lambda + 2 \frac{1}{3 + 2\theta} = 0 .$$

By the Jury conditions, it is easy to obtain $R_0 > 1$ for all $\theta > 0$. This means the population dispersal diffuses the disease spread.

Let us now fix $r = c = \gamma = 1, \mu - 2, \beta_2 = 1, \beta_1 = 4$. Then $R_{01} = 4/3, R_{02} < 1/3$. Now, the characteristic equation becomes

$$\lambda^2 - \frac{5}{3} \frac{3+\theta}{3 + 2\theta} \lambda + \frac{4}{3} \frac{1}{3 + 2\theta} = 0 .$$

By the Jury conditions, we have $R_0 > 1$ if $0 \leq \theta < 2$, and $R_0 < 1$ if $\theta > 2$. Thus, increasing population dispersal can reduce the risk of the disease spread.

Example 2. Fix $r_1 = 1, r_2 = 5$ and $\beta_1 = 1.5, \beta_2 = 0.1$. This means that we have different birth coefficients and different disease transmission coefficients in different patches. Let us choose $c_1 = c_2 = 1, \mu_1 = \mu_2 = 2, \gamma_1 = \gamma_2 = 0$. Then $R_{01} = 0.75, R_{02} = 0.25$. Thus, the disease dies out in each patch when two patches are uncoupled. Now, we fix $a_{11} = b_{11} = 0.3, a_{22} = b_{22} = 0.3k$, where k is a positive constant. This means susceptible individuals and infected individuals in each patch have the same dispersal rate, but different patches may have different migration rates. By direct calculations, we have $R_0 < 1$ when $0 < k < 0.7138728627, R_0 > 1$ when $k > 0.7138728627$. Thus, the disease will spread in the two patches if $k > 0.7138728627$, although the disease can not spread in any patch when they are isolated.

If infectives are barred at borders, that is, $b_{11} = b_{22} = 0$, we have

Theorem 3 (Wang 2004). *Assume that $a_{11} > 0, a_{22} > 0$ and $\gamma_1 = \gamma_2 = b_{11} = b_{22} = 0$. If $R_{01} < 1$ and $R_{02} < 1$, the disease either disappears in the two patches or spreads in one patch and dies out in the other patch except in certain critical cases.*

This theorem means that the disease may spread in one patch even though movements of infected individuals between two patches are not allowed.

Example 3. Assume $\gamma_1 = \gamma_2 = b_{11} = b_{22} = 0$. Fix $r_1 = 1, \mu_1 = 1, c_1 = 0.5, \beta_1 = 0.4, r_2 = 1, \mu_2 = 1, c_2 = 0.4, \beta_2 = 0.55$. Thus, infected individuals can not pass through borders of the patches. Then $R_{01} = 0.8, R_{02} = 0.9166666668$. Now, we choose $a_{11} = 0.4$ and set

$$R_0 = \max\left\{0.4\,\frac{0.6 + 2\,a_{22}}{0.54 + 0.5\,a_{22}}, 0.715\,(0.54 + 0.5\,a_{22})^{-1}\right\}.$$

As a_{22} increases from 0, we have $R_0 > 1$ if $0 < a_{22} < 0.35$, $R_0 < 1$ if $0.35 < a_{22} < 1$, and $R_0 > 1$ if $a_{22} > 1$. By similar arguments as those in Wang and Zhao (2004), we see that the disease spreads in the second patch and dies out in the first patch if $0 < a_{22} < 0.35$, the disease dies out in two patches if $0.35 < a_{22} < 1$, and the disease spreads in the first patch and dies out in the second patch if $a_{22} > 1$. Therefore, population movements can intensify a disease spread and the migration from the second patch to the first patch reduces the risk of a disease outbreak in the second patch.

Theorem 4 (Wang 2004). *Assume that $b_{11} = b_{22} = 0$. If $R_{01} > 1$ and $R_{02} > 1$, the disease either spreads in the two patches or spreads in one patch and dies out in the other patch except in certain critical cases.*

4.5 A constant size with a standard incidence

In this section, we assume that the birth rates and death rates in each patch are equal and we change the bilinear incidence to a standard incidence, which is often the case in epidemiology. We suppose that the dynamics of the individuals are governed by

$$\begin{aligned}
\frac{dS_1}{dt} &= \mu_1 N_1 - \mu_1 S_1 - \beta_1 S_1 \frac{I_1}{N_1} + \gamma_1 I_1 - a_1 S_1 + a_2 S_2\,, \\
\frac{dS_2}{dt} &= \mu_2 N_2 - \mu_2 S_2 - \beta_2 S_2 \frac{I_2}{N_2} + \gamma_2 I_2 + a_1 S_1 - a_2 S_2\,, \\
\frac{dI_1}{dt} &= \beta_1 S_1 \frac{I_1}{N_1} - (\mu_1 + \gamma_1)I_1 - b_1 I_1 + b_2 I_2\,, \\
\frac{dI_2}{dt} &= \beta_2 S_2 \frac{I_2}{N_2} - (\mu_2 + \gamma_2)I_2 - b_2 I_2 + b_1 I_1\,,
\end{aligned} \tag{21}$$

where a_1 represents the rate at which susceptible individuals migrate from the first patch to the second patch, a_2 is the rate at which susceptible individuals migrate from the second patch to the first patch, b_1 is the rate at which infectious individuals migrate from the first patch to the second patch, b_2 is the rate at which infected individuals migrate from the second patch to the first patch. In this model, we neglect the death and birth processes of the individuals when they are dispersing and suppose that $\mu_i, \beta_i, i = 1, 2$, are positive constants, $a_i, b_i, i = 1, 2$, are nonnegative constants.

If $N = N_1 + N_2 = S_1 + I_1 + S_2 + I_2$, it follows from $N' = 0$ that N is a constant. In what follows, we suppose $N(t) \equiv A > 0$. Set $s_1 = S_1/N_1, i_1 = I_1/N_1, s_2 = S_2/N_2, i_2 = I_2/N_2$ and define $x = 1/N_1$. By direct calculations, we see that (21) can be reduced to

$$
\frac{\mathrm{d}i_1}{\mathrm{d}t} = (a_1 + a_2 + \beta_1 - b_1 - \mu_1 - \gamma_1)i_1 + (b_1 - a_1 - \beta_1)i_1^2
$$
$$
- b_2 i_2 + (b_2 - a_2)i_1 i_2 + A\,x\,(b_2\,i_2 - a_2 i_1 + (a_2 - b_2)i_1 i_2)\,,
$$

$$
\frac{\mathrm{d}i_2}{\mathrm{d}t} = \frac{1}{Ax - 1}\{b_1 i_1 + (b_2 + \mu_2 + \gamma_2 - a_1 - a_2 - \beta_2)i_2
$$
$$
+ (a_2 + \beta_2 - b_2)i_2^2 + (a_1 - b_1)i_1 i_2 \tag{22}
$$
$$
- Axi_2[(\beta_2 + a_2 - b_2)i_2 - \beta_2 - a_2 + \mu_2 + \gamma_2 + b_2]\}\,,
$$

$$
\frac{\mathrm{d}x}{\mathrm{d}t} = x[a_1 + a_2 + (b_1 - a_1)i_1 + (b_2 - a_2)i_2 - A\,x(a_2 + (b_2 - a_2)i_2)]\,.
$$

Clearly, a reasonable region for this model is

$$
X = \{(i_1, i_2, x)\colon 0 \le i_1 \le 1, 0 \le i_2 \le 1, 1/A < x\}\,.
$$

Define

$$
F = \begin{bmatrix} \beta_1 & 0 \\ 0 & \beta_2 \end{bmatrix}
$$

$$
\mathcal{V} = -\begin{bmatrix} -\mu_1 - \gamma_1 - b_1 & b_2 \\ b_1 & -\mu_2 - \gamma_2 - b_2 \end{bmatrix}\,.
$$

If $R_0 := \rho(F\mathcal{V}^{-1})$, where $\rho(F\mathcal{V}^{-1})$ is the spectral radius of matrix $F\mathcal{V}^{-1}$, as discussed above, we see that R_0 is a basic reproduction number of (21).

$E_0 = (0, 0, (a_1 + a_2)/(A\,a_2))$ is a disease free equilibrium. By Wang and Giuseppe (2003), we have the following results:

Theorem 5. *Let $a_1 > 0, a_2 > 0$. Then the disease-free equilibrium of (22) is globally stable if $R_0 < 1$.*

Theorem 6. *Let $b_2 = 0$ or $b_1 = 0$ and let $a_i \ge 0, i = 1, 2$. Then for any solution of (21) in X, we have $\lim\limits_{t \to \infty} I_j(t) = 0, j = 1, 2$, i.e., the disease becomes extinct in the two patches, if*

$$
\begin{cases} \beta_1 < \mu_1 + \gamma_1 + b_1\,, \\ \beta_2 < \mu_2 + \gamma_2 + b_2\,. \end{cases} \tag{23}
$$

Theorem 7. *Let $a_i > 0$ and $b_i > 0, i = 1, 2$. If $R_0 > 1$, then the disease is uniformly persistent in the two patches, i. e., there is a positive constant ϵ such that every positive solution $(i_1(t), i_2(t), x(t))$ of (22) satisfies*

$$\liminf_{t \to \infty} i_1(t) \geq \epsilon, \quad \liminf_{t \to \infty} i_2(t) \geq \epsilon.$$

Theorem 8. *Let $b_2 = 0, b_1 > 0, a_1 > 0, a_2 > 0$. Then there is a positive constant ϵ such that for any positive solution $(S_1(t), S_2(t), I_1(t), I_2(t))$ of (21), we have $\liminf_{t \to \infty} I_j(t) > \epsilon, j = 1, 2$, i. e., the disease is uniformly persistent in the two patches, if*

$$\frac{\beta_1}{\mu_1 + \gamma_1 + b_1} > 1. \tag{24}$$

Theorem 9. *There is a positive constant ϵ such that for any positive solution $(S_1(t), I_1(t), S_2(t), I_2(t))$ of (21), we have $\liminf_{t \to \infty} I_j(t) > \epsilon, j = 1, 2$, if one of the following conditions is satisfied:*

(i) $b_1 = 0, b_2 > 0, a_1 > 0, a_2 > 0$ and $\frac{\beta_2}{\mu_2 + \gamma_2 + b_2} > 1$,
(ii) $b_1 = 0, b_2 = 0, a_1 > 0, a_2 > 0$ and $\beta_i > \mu_i + \gamma_i, i = 1, 2$.

Let us discuss the implications of Theorems 5–9.

Remark 1. Theorems 5–9 show that the threshold conditions for the spread of the disease are independent of a_1 and a_2. Thus, the dispersal rates of susceptible individuals do not influence the permanence of the disease. In contrast, the dispersal rates of susceptible individuals play key roles to the outbreak of a disease when the incidences obey a mass action law.

Remark 2. If two patches are isolated, it is easy to see that the disease spreads in the i-th patch if $\beta_i > \mu_i + \gamma_i$ and the disease becomes extinct if $\beta_i < \mu_i + \gamma_i$. Let us now consider the influence of population dispersal on disease spread when $a_1 > 0, a_2 > 0$. First, we suppose that $\beta_i < \mu_i + \gamma_i, i = 1, 2$, i. e., the disease becomes extinct in each patch when they are isolated. It is easy to obtain a characteristic equation for matrix FV^{-1}:

$$\lambda^2 - q_1 \lambda + q_2 = 0,$$

where

$$q_1 = \frac{\beta_1 \mu_2 + \beta_1 \gamma_2 + \beta_1 b_2 + \beta_2 \mu_1 + \beta_2 \gamma_1 + \beta_2 b_1}{\mu_1 \mu_2 + \mu_1 \gamma_2 + \mu_1 b_2 + \gamma_1 \mu_2 + \gamma_1 \gamma_2 + \gamma_1 b_2 + b_1 \mu_2 + b_1 \gamma_2},$$

$$q_2 = \frac{\beta_1 \beta_2}{\mu_1 \mu_2 + \mu_1 \gamma_2 + \mu_1 b_2 + \gamma_1 \mu_2 + \gamma_1 \gamma_2 + \gamma_1 b_2 + b_1 \mu_2 + b_1 \gamma_2}.$$

Using the conditions that $\beta_i < \mu_i + \gamma_i, i = 1, 2$, we can verify the Jury conditions (20) are satisfied. Thus, $R_0 < 1$. Hence, if the disease becomes extinct when the two patches are isolated, the disease remains extinct when population dispersal occurs.

Now, we consider the case where $\beta_i > \mu_i + \gamma_i, i = 1, 2$, i. e., the disease spreads in each patch when they are isolated. Then Theorems 7–9 show that the disease remains permanent when the two patches are connected. In fact, if $b_1 = 0$ or $b_2 = 0$, it follows from Theorem 8 or Theorem 9 that the disease is uniformly persistent in the two patches. Suppose $b_1 > 0$ and $b_2 > 0$. Then by checking the Jury conditions, we have $R_0 > 1$. It follows from Theorem 7 that the disease spreads in the two patches.

We can also discuss the stability of an endemic equilibrium of (22) since it is a three dimensional system. First, we suppose that the susceptible individuals of the two patches disperse between the two patches but the infective individuals do not. This means that we suppose that $a_1 > 0, a_2 > 0, b_1 = 0$ and $b_2 = 0$. Suppose that

$$\beta_1 > \mu_1 + \gamma_1 \,, \quad \beta_2 > \mu_2 + \gamma_2 \,. \tag{25}$$

Then (22) has a unique positive equilibrium (i_1^*, i_2^*, x^*) where

$$i_1^* = \frac{\beta_1 - \mu_1 - \gamma_1}{\beta_1} \,,$$

$$i_2^* = \frac{\beta_2 - \mu_2 - \gamma_2}{\beta_2} \,,$$

$$x^* = \frac{a_1\,\beta_2\,\gamma_1 + a_1\,\beta_2\,\mu_1 + a_2\,\beta_1\,\mu_2 + a_2\,\beta_1\,\gamma_2}{\beta_1\,A\,a_2\,(\mu_2 + \gamma_2)} \,.$$

By Routh-Hurwitz criteria and by means of Maple, we have:

Theorem 10. *Suppose (25) holds. Then the positive equilibrium is asymptotically stable.*

Next, we suppose that susceptible individuals have the same dispersal rate as infectious individuals in each patch, i. e., $a_i = b_i, i = 1, 2$. Then system (21) becomes

$$\frac{dS_1}{dt} = \mu_1 N_1 - \mu_1 S_1 - \beta_1 S_1 \frac{I_1}{N_1} + \gamma_1 I_1 - a_1 S_1 + a_2 S_2 \,,$$

$$\frac{dS_2}{dt} = \mu_2 N_2 - \mu_2 S_2 - \beta_2 S_2 \frac{I_2}{N_2} + \gamma_2 I_2 + a_1 S_1 - a_2 S_2 \,,$$

$$\frac{dI_1}{dt} = \beta_1 S_1 \frac{I_1}{N_1} - (\mu_1 + \gamma_1)I_1 - a_1 I_1 + a_2 I_2 \,, \tag{26}$$

$$\frac{dI_2}{dt} = \beta_2 S_2 \frac{I_2}{N_2} - (\mu_2 + \gamma_2)I_2 - a_2 I_2 + a_1 I_1 \,.$$

Theorem 11. *Let $a_1 > 0$ and $a_2 > 0$. Suppose $R_0 > 1$. Then each positive solution $(S_1(t), S_2(t), I_1(t), I_2(t))$ of (26) converges to a positive steady state as t tends to infinity, i. e., there is an $(S_1^*, S_2^*, I_1^*, I_2^*) > 0$ such that*

$$\lim_{t \to \infty} (S_1(t), S_2(t), I_1(t), I_2(t)) = (S_1^*, S_2^*, I_1^*, I_2^*) \,.$$

In the general case except for the above two cases, the stability of an endemic equilibrium is open.

4.6 Patch models with differentiating residence

In the last section, we have included migrations and emigrations of individuals into the epidemic models. It is assumed that an individual who moves to a new patch will become a resident of the new patch. Sattenspiel and Dietz (1995) proposed an epidemic model with geographic mobility among regions in which a person does not change its residence during movements, which is presented below.

4.6.1 Mobility model

Consider a population distributed into n patches. Individuals from region i leave on trips to other regions at a per capita rate σ_i per unit time. These visitors are distributed among the $n-1$ destinations with probabilities ν_{ij} to each destination j. Because the ν_{ij} give conditional probabilities of visiting another region, we have $0 \leq \nu_{ij} \leq 1$ for $i \neq j$ and, by definition, $\nu_{ii} = 0$. Furthermore, $\sum_{j=1}^{k} \mu_{ij} = 1$. Persons travelling from patch i to patch j have a per capita return rate to region i of ρ_{ij}. By definition, $\rho_{ii} = 0$.

Let $N_{ii}(t)$ be the number of residents of region i who are actually present in their home region at time t, and let $N_{ij}(t)$ be the number of residents of region i who are visiting region j at time t. Then the travel patterns of individuals among regions lead to the equations

$$\frac{\mathrm{d}N_{ii}}{\mathrm{d}t} = \sum_{j=1}^{k} \rho_{ij} N_{ij} - \sigma_i N_{ii} \,,$$

$$\frac{\mathrm{d}N_{ij}}{\mathrm{d}t} = \sigma_i \nu_{ij} N_{ii} - \rho_{ij} N_{ij} \,. \tag{27}$$

4.6.2 An epidemic model with mobility

Transmission of an infectious agent in a mobile population requires that the following events occur: (1) A susceptible person travels from her home patch i to some patch k, (2) An infective person travels from his home patch to the same patch k, (3) Contact occurs among people at patch k, and in some proportion of the contacts between a susceptible person and an infectious person the infectious organism is transmitted.

Assume that κ_k is the average number of contacts per person made in region k, β_{ijk} is the proportion of contacts in region k between a susceptible from region i and an infective from region j that actually result in transmission of the infection, I_{jk} is the number of infectives present in region k who are permanent residents of region j, S_{ik} is the number of susceptibles present in region k who are permanent residents of region i, and $N_k^* = \sum_{m=1}^{k}(S_{mk} + I_{mk} + R_{mk})$ is the number of people actually present in region k. In principle, the number of contacts per person could be a function of the home locations of the people involved, which might be important if, say, cultural background influenced how gregarious a person was. For simplicity's sake, it is assumed in paper (Sattenspiel and Dietz 1995) that it is a simple function of location of contact only.

If we consider an SIR epidemic disease, the equations for change in number of susceptible residents of patch i who are actually present in that region is derived as follows:

$$
\begin{aligned}
\frac{\mathrm{d}S_{ii}}{\mathrm{d}t} = {} & \text{no. of residents returning home} \\
& - \text{no. of residents leaving on trips} - \text{new transmissions} \\
= {} & \sum_{k=1}^{n} \rho_{ik} S_{ik} - \sigma_i S_{ii} - \sum_{j=1}^{n} \kappa_i \beta_{iji} \frac{S_{ii} I_{ji}}{N_i^*} \ .
\end{aligned}
\tag{28}
$$

The equations for susceptible residents of patch i who are visiting other patches are derived similarly and are given by

$$
\frac{\mathrm{d}S_{ik}}{\mathrm{d}t} = \sigma_i \nu_{ik} S_{ii} - \rho_{ik} S_{ik} - \sum_{j=1}^{n} \kappa_k \beta_{ijk} \frac{S_{ik} I_{jk}}{N_k^*} \ .
\tag{29}
$$

Summing these equations for all patches gives

$$
\frac{\mathrm{d}S_i}{\mathrm{d}t} = -\sum_{k=1}^{n}\sum_{j=1}^{n} \kappa_k \beta_{ijk} \frac{S_{ik} I_{jk}}{N_k^*}
$$

where S_i is the total number of residents from patch i. It implies that S_i is altered only through disease transmission and not as a consequence of the mobility process. Thus, the model does not include permanent migration.

Equations for the other disease classes are derived similarly, and are given by

$$
\begin{aligned}
\frac{\mathrm{d}I_{ii}}{\mathrm{d}t} &= \sum_{k=1}^{n} \rho_{ik} I_{ik} - \sigma_i I_{ii} + \sum_{j=1}^{n} \kappa_i \beta_{iji} \frac{S_{ii} I_{ji}}{N_i^*} - \gamma I_{ii} \ , \\
\frac{\mathrm{d}I_{ik}}{\mathrm{d}t} &= \sigma_i \nu_{ik} I_{ii} - \rho_{ik} I_{ik} + \sum_{j=1}^{n} \kappa_k \beta_{ijk} \frac{S_{ik} I_{jk}}{N_k^*} - \gamma I_{ik} \ , \\
\frac{\mathrm{d}R_{ii}}{\mathrm{d}t} &= \sum_{k=1}^{n} \rho_{ik} R_{ik} - \sigma_i R_{ii} + \gamma I_{ii} \ , \\
\frac{\mathrm{d}R_{ik}}{\mathrm{d}t} &= \sigma_i \nu_{ik} R_{ii} - \rho_{ik} R_{ik} + \gamma I_{ik} \ ,
\end{aligned}
\tag{30}
$$

where γ is the rate of recovery from the disease.

The general structure of the mobility process can be used in models for a number of different kinds of diseases and can be extended to allow for multiple mobility patterns (see paper (Sattenspiel and Dietz 1995) for further details).

4.7 Models with residence and demographic structure

In the model introduced in paper (Sattenspiel and Dietz 1995) there is no intra-patch demography (no birth or natural death of individuals), only inter-patch travel. Arino and van den Driessche improved this in papers (Arino 2003a; Arino and Van den Driessche 2003b).

4.7.1 Mobility model of population

To make the model a little more realistic, but in order to work with a constant overall population, it is assumed in Arino (2003a) that birth and death occur with the same rate. In addition, it is assumed that individuals who are out of their home patch do not give birth, and so birth occurs in the home patch at a per capita rate $d > 0$, and death takes place anywhere with a per capita rate d.

Suppose that the total number of patches is n. In the following, we call residents of a patch i the individuals who were born in and normally live in that patch, and travellers the individuals who at the time they are considered, are not in the patch they reside in. We denote the number of residents of patch i who are present in patch j at time t by N_{ij}. Letting N_i^r be the resident population of patch i at time t, then

$$N_i^r = \sum_{j=1}^{n} N_{ij} \,.$$

Also, letting N_i^p be the population of patch i at time t, i. e., the number of individuals who are physically present in patch i, both residents and travellers, then

$$N_i^p = \sum_{j=1}^{n} N_{ji} \,.$$

Let g_i be the rate at which residents of patch i leave their home patch, $m_{ji} \geq 0$ be the fraction of these outgoing individuals who go to patch j. Thus, if $g_i > 0$, then $\sum_{j=1}^{n} m_{ji} = 1$, with $m_{ii} = 0$, and $g_i m_{ji}$ is the travel rate from patch i to patch j. Residents of patch i who are in patch j return to patch i with a per capita rate $r_{ij} \geq 0$, with $r_{ii} = 0$. These assumptions imply that an individual resident in a given patch, say patch i, who is present in some patch j, must first return to patch i before travelling to another patch k, where i, j, k are distinct.

Since birth occurs in the home patch and death takes place anywhere, from the above assumptions we have

$$\frac{\mathrm{d}N_{ii}}{\mathrm{d}t} = d(N_i^r - N_{ii}) + \sum_{j=1}^{n} r_{ij} N_{ij} - g_i N_{ii} \,. \tag{31}$$

The evolution of the number of residents of patch i who are present in patch $j, j \neq i$, is

$$\frac{\mathrm{d}N_{ij}}{\mathrm{d}t} = g_i m_{ji} N_{ii} - r_{ij} N_{ij} - dN_{ij} \, . \tag{32}$$

By (31) and (32), we have $\frac{\mathrm{d}N_i^r}{\mathrm{d}t} = 0$. Thus, the number of residents of patch i is a fixed quantity. As a consequence, the overall size of the population in n patches is a constant. However, the number of individuals present in patch i is in general a variable quantity.

Equations (31) and (32) constitute the mobility model. Since it is linear, it is not hard to see that the model has a globally asymptotically stable equilibrium $(\widehat{N}_{11}, \ldots, \widehat{N}_{nn})$ where

$$\widehat{N}_{ii} = \left(\frac{1}{1 + g_i C_i}\right) N_i^r \, , \quad \widehat{N}_{ij} = g_i \frac{m_{ji}}{d + r_{ij}} \left(\frac{1}{1 + g_i C_i}\right) N_i^r \tag{33}$$

with $C_i = \sum_{k=1}^n \frac{m_{ki}}{d + r_{ik}}$ for $i = 1, \ldots, n$.

4.7.2 Epidemic model with mobility

Based upon the model of population mobility in the last subsection, paper (Arino 2003a) proposed an SIS model.

Let S_{ij} and I_{ij} denote the number of susceptible and infective individuals resident in patch i who are present in patch j at time t; thus $N_{ij} = S_{ij} + I_{ij}$ for all $i, j = 1, \ldots, n$. Disease transmission is modelled using standard incidence, which, for human diseases, is considered more accurate than mass action (see, e. g., Hethcote (2000); McCallum et al. 2001). In patch j, this gives

$$\sum_{k=1}^n \kappa_j \beta_{ikj} \frac{S_{ij} I_{kj}}{N_j^p}$$

where the disease transmission coefficient $\beta_{ikj} > 0$ is the proportion of adequate contacts in patch j between a susceptible from patch i and an infective from patch k that actually results in transmission of the disease and $\kappa_j > 0$ is the average number of such contacts in patch j per unit time. Let $\gamma > 0$ denote the recovery rate of infectives, thus $1/\gamma$ is the average infective period. Note that γ is assumed to be the same for all cities.

In each patch, there are $2n$ equations. The first n equations describe the dynamics of the susceptibles, and the n others describe the dynamics of the infectives. Since there are n patches, there is a total of $2n^2$ equations for n patches. The dynamics of the number of susceptibles and infectives originating from patch i (with $i = 1, \ldots, n$) is given by the following system:

$$\frac{\mathrm{d}S_{ii}}{\mathrm{d}t} = \sum_{k=1}^n r_{ik} S_{ik} - g_i S_{ii} - \sum_{k=1}^n \kappa_i \beta_{iki} \frac{S_{ii} I_{ki}}{N_i^p} + d(N_i^r - S_{ii}) + \gamma I_{ii}$$

$$\frac{\mathrm{d}I_{ii}}{\mathrm{d}t} = \sum_{k=1}^n r_{ik} I_{ik} - g_i I_{ii} + \sum_{k=1}^n \kappa_i \beta_{iki} \frac{S_{ii} I_{ki}}{N_i^p} - (\gamma + d) I_{ii} \, , \tag{34}$$

and, for $j \neq i$,

$$\frac{\mathrm{d}S_{ij}}{\mathrm{d}t} = g_i m_{ji} S_{ii} - r_{ij} S_{ij} - \sum_{k=1}^{n} \kappa_j \beta_{ikj} \frac{S_{ij} I_{kj}}{N_j^p} - dS_{ij} + \gamma I_{ij}$$

$$\frac{\mathrm{d}I_{ij}}{\mathrm{d}t} = g_i m_{ji} I_{ii} - r_{ij} I_{ij} + \sum_{k=1}^{n} \kappa_j \beta_{ikj} \frac{S_{ij} I_{kj}}{N_j^p} - (d + \gamma) I_{ij} \,. \tag{35}$$

The system of (34) and (35) is at the disease free equilibrium if $I_{ji} = 0$ and $S_{ji} = \widehat{N}_{ji}$ given by (33) for all $i, j = 1, \ldots, n$. To obtain the basic reproduction number of the model of (34) and (35), we order the infective variables as

$$I_{11}, \ldots, I_{1n}, I_{21}, \ldots, I_{2n}, \ldots, I_{nn} \,.$$

Then we define a diagonal block matrix $V = \mathrm{diag}(V_{ii})$ for $i = 1, \ldots, n$, where

$$V_{ii} = \begin{bmatrix} r_{i1} + \gamma + d & 0 & \cdots & -g_i m_{1i} & 0 & \cdots & 0 \\ 0 & r_{i2} + \gamma + d & \cdots & -g_i m_{2i} & 0 & \cdots & 0 \\ \cdots & \cdots & \cdots & \cdots & \cdots & \cdots & \cdots \\ -r_{i1} & -r_{i2} & \cdots & g_i + \gamma + d & 0 & \cdots & -r_{in} \\ \cdots & \cdots & \cdots & \cdots & \cdots & \cdots & \cdots \\ 0 & \cdots & \cdots & -g_i m_{ni} & 0 & \cdots & r_{in} + \gamma + d \end{bmatrix} \,.$$

It is not hard to verify that V_{ii} is a nonsingular M-matrix and the inverse of V is a nonnegative matrix $V^{-1} = \mathrm{diag}(V_{ii}^{-1})$. Next, we define a block matrix $F = (F_{ij})$ where each block F_{ij} is $n \times n$ diagonal and has the form $F_{ij} = \mathrm{diag}(f_{ijq})$ with

$$f_{ijq} = \kappa_q \beta_{ijq} \frac{\widehat{N}_{iq}}{\widehat{N}_q^p}$$

for $q = 1, \ldots, n$. By Diekmann et al. (1990) and van den Driessche and Watmough (2002),

$$R_0 := \rho(FV^{-1}) \,,$$

is the basic reproduction number for the model of (34) and (35), where ρ represents the spectral radius of the matrix. It can shown that the disease free equilibrium is locally asymptotically stable if $R_0 < 1$ and is unstable if $R_0 > 1$.

Arino and van den Driessche (2003a) found the following interesting phenomena. Consider two patches. Suppose that $N_1^r = N_2^r = 1500, d = 1/(75 \times 365), \gamma = 1/25, \kappa_1 = \kappa_2 = 1$ and $r_{12} = r_{21} = 0.05, \beta_1 \approx 0.016, \beta_2 \approx 0.048$. If $g_1 = 0.4$ and g_2 is increased, then there are two successive bifurcations. For small g_2, there is a unique endemic equilibrium; for intermediate g_2, there is no endemic equilibrium and the disease dies out; for large g_2 a unique endemic equilibrium is again present. Also if the rate of leaving is the same in each patch ($g_1 = g_2$), then two bifurcations are also observed, with an endemic equilibrium present for large g_i. These illustrate the complexity of behavior possible when inter-patch travel is present.

4.8 Discussion

The majority of models of the dynamics of infectious diseases have assumed the existence of populations with homogeneous mixing. These models suppose that all individuals have ecological and epidemiological structures that are independent of space. It has been increasingly recognized that space plays an important role in many infectious disease processes because populations are not well-mixed: interactions between individuals tend to be mainly local in nature and disease incidence records clearly illustrate non-uniformities in the spatial distribution of cases. In this chapter, we have presented recent advances of epidemic models with geographic effects. The main points are to introduce metapopulation theory (see Hanski (1999); Levin 1974; Okubo and Levin 2001; Takeuchi and Lu (1986, 1992)) into classical epidemic models.

The epidemic model by Brauer and van den Driessche (2001) is used to simulate phenomena that travellers may return home from a foreign trip with an infection acquired abroad. This should be the simplest way to include a population mobility. Since there is a fixed input of infectives from outside, the disease is always persistent. Lloyd et al. (1996, 2004) proposed multi-patch models for epidemic diseases. These models have included spatial heterogeneity, but assumed that each patch has a constant population size although persons move between the patches, and imply that mobile persons are visitors and stay in other patches in very short period so that only epidemiological cross-coupling between patches is considered. For a simplified case where each patch is of the same population size and the contact rate is the same within each patch, paper (Lloyd et al. 1996) found that the endemic equilibrium is stable and the patches often oscillate in phase for all but the weakest between patch coupling. Rodríguez and Torres-Sorando (2001) proposed a malaria model with spatial heterogeneity. It is assumed that each patch has constant sizes of human persons and mosquitoes, further, humans can travel between patches but mosquitoes can not. Under the assumptions that patches are equal in a sense that humans and mosquitoes have the same sizes in every patch, paper (Rodríguez and Torres-Sorando 2001) obtained conditions (8) and (9) for the invasion of the malaria. These conditions show that the establishment of the disease is more restricted as the environment is more partitioned.

We have incorporated demographic structures into multi-patch epidemic models in papers (Wang 2004; Wang and Mulone 2003; Wang and Zhao (2004)). Our models have relaxed the assumption that population size in each patch is constant. Thus, the timescale of an epidemic disease is allowed to be longer and individuals can migrate from one patch to other patches. We have obtained the basic reproduction number of model (12) by using the next generation matrix concept from Diekmann et al. (1990) and van den Driessche, Watmough (Van den Driessche and Watmough 2002). Theorem 1 means that the disease free equilibrium is asymptotically stable if the basic reproduction number is less than 1. Theorem 2 shows that the disease is uniformly persistent if the basic reproduction number is greater than 1. Thus, the basic reproduction number is a threshold for the invasion of the disease. For the special case where there are only two patches, if two patches have the same demographic parameters, we have shown that $R_{01} < 1$ and $R_{02} < 1$ imply $R_0 < 1$, $R_{01} > 1$ and $R_{02} > 1$ imply $R_0 > 1$. Thus, the disease is uniformly persistent if it is uniformly persistent in each disconnected patch. We have also shown that population dispersal can facilitate a disease spread or reduce a disease spread if $R_{01} < 1$ and $R_{02} > 1$. Example 2 shows that disease can spread in the two patches even though the disease can not spread in any patch when they are disconnected. Theorem 3 means that the disease may spread in one patch even though movements of infected individuals between two patches are barred. For the model (21) where standard incidences are adopted and the number of patches is 2, Theorems 5–9 show that the threshold conditions for the spread of the disease are independent of a_1 and a_2. Thus, the dispersal rates of susceptible individuals do not influence the permanence of the disease. In contrast, the dispersal rates of susceptible individuals play key roles to the outbreak of a disease when the incidences obey a mass action law.

Sattenspiel and Dietz (1995) formulated the epidemic model of (28), (29) and (30) with geographic mobility among regions in which a person does not change its residence during movements. But there is no intra-patch demography (no birth or natural death of individuals) in that model. Arino and van den Driessche improved this in the model of (34) and (35). But it is assumed that individuals do not give birth when they are not in their home patches, and an individual resident in a given patch, say patch i, who is present in some patch j, must first return to home patch i before travelling to another patch k, where i, j, k are distinct. Arino and van den Driessche (2003a) found that as a dispersal rate g_2 increases, if its values are small, there is a unique endemic equilibrium; for intermediate g_2, there is no endemic equilibrium and the disease dies out; for large g_2 a unique endemic equilibrium is again present. This illustrates the complexity of behavior possible when inter-patch travel is present.

Multi-patch epidemic models have been used to analyze dynamics of specific diseases. See Fulford et al. (2002) for tuberculosis in possums, Grenfell and Harwood (1997), Keeling and Gilligan (2000) for bubonic plague, Sattenspiel and Herring (2003) for influenza.

References

1. J. Arino; P. van den Driessche, A multi-city epidemic model. *Math. Popul. Stud.* **10** (2003a), 175–193.
2. Arino, J. and P. van den Driessche (2003b), The basic reproduction number in a multi-city compartmental epidemic model. Positive systems (Rome), 135–142, Lecture Notes in Control and Inform. Sci., 294, Springer, Berlin.

3. Brauer, F. and P. van den Driessche (2001), Models for transmission of disease with immigration of infectives. *Math. Biosci*, **171**, 143–154.
4. Britton, N. F. (2003), *Essential Mathematical Biology*. Springer-Verlag London, Ltd., London.
5. Cooke, K. L. and P. van den Driessche (1996), Analysis of an SEIRS epidemic model with two delays. *J. Math. Biol.* **35**, 240–260.
6. Cooke, K., P. van den Driessche and X. Zou (1999), Interaction of maturation delay and nonlinear birth in population and epidemic models. *J. Math. Biol.* **39**, 332–352.
7. Diekmann, O., J.A.P. Heesterbeek and J.A.J. Metz (1990), On the definition and the computation of the basic reproduction ratio R_0 in the models for infectious disease in heterogeneous populations. *J. Math. Biol.* **28** , 365–382.
8. Diekmann, O. and J. A. P. Heesterbeek (2000), *Mathematical epidemiology of infectious diseases. Model building, analysis and interpretation*. Wiley Series in Mathematical and Computational Biology. John Wiley & Sons, Ltd., Chichester.
9. Feng, Z. and H.R. Thieme (2000a), Endemic modls with arbitrary distributed periods of infection I: fundamental properties of the model. *SIAM J.Appl. Math.* **61**, 803–833.
10. Feng, Z. and H.R. Thieme (2000b), Endemic modls with arbitrary distributed periods of infection II: fast disease dynamics and permanent recovery. *SIAM J.Appl. Math.* **61**, 983–1012.
11. Freedman, H. I. and P. Waltman (1984), Persistence in models of three interacting predator-prey populations. *Math. Biosci.* **68**, 213–231.
12. Fulford, G. R., M. G. Roberts and J. A. P. Heesterbeek (2002), The metapopulation dynamics of an infectious disease: tuberculosis in possums, *Theoretical Population Biology*. **61**, 15–29.
13. Gyllenberg, M. and G. F. Webb (1990), A nonlinear structured population model of tumor growth with quiescence. *J. Math. Biol.* **28**, 671–694.
14. Grenfell, B. and J. Harwood (1997), (Meta)population dynamics of infectious diseases. *Trends in Ecology and Evolution.* **12**, 395–399.
15. Han, L., Z. Ma, S. Tan (2003), An SIRS epidemic model of two competitive species. *Math. Comput. Modelling* **37**, 87–108.
16. Hethcote, H.W. (2000), The mathematics of infectious diseases. *SIAM Review.* 42, 599–653.
17. Hanski, I.(1999), *Metapopulation ecology*. Oxford University Pree.
18. Inaba, H. (1990), Threshold and stability results for an age-structured epidemic model. *J. Math. Biol.* **28**, 411–434.
19. M. J. Keeling and C. A. Gilligan (2000), Metapopulation dynamics of bubonic plague. *Nature.* **407**, 903–906.
20. Lajmanovich and Yorke (1976), A deterministic model for gonorrhea in a nonhomogeneous population. *Math. Biosci.* **28**, 221–236.
21. Levin, S. A. (1974), Dispersion and population interactions. Amer. Natur. **108**, 207–228.
22. Liu, W. M., S. A. Levin and Y. Iwasa (1986), Influence of nonlinear incidence rates upon the behavior of SIRS epidemiological models, *J. Math. Biol.* **23**, 187–204.
23. Liu, W. M., H. W. Hethcote and S. A. Levin (1987), Dynamical behavior of epidemiological models with nonlinear incidence rates, *J. Math. Biol.* **25**, 359–380.

24. Lloyd, A. L. and R. M. May (1996), Spatial heterogeneity in epidemic models. *J. Theor. Biol.* **91**, 1–11.

25. Lloyd, A. L. and V. A. A. Jansen (2004), Spatiotemporal dynamics of epidemics: synchrony in metapopulation models. *Math. Biosci.* **188**, 1–16.

26. Lu, Y. and Y. Takeuchi (1992), Permanence and global stability for cooperative Lotka-Volterra diffusion systems. *Nonlinear Anal.* **19**, 963–975.

27. Ma, W., Y. Takeuchi, T. Hara and E. Beretta (2002), Permanence of an SIR epidemic model with distributed time delays. *Tohoku Math. J.* **54**, 581–591.

28. McCallum, H., N. Barlow and J. Hone (2001), How should pathogen transmission be modelled?. *Trends Ecol. Evol.* **16**, 295–300.

29. Murray, J. D. (1989), *Mathematical biology.* Springer-Verlag, Berlin.

30. Okubo, A. and S. A. Levin (2001), *Diffusion and ecological problems: modern perspectives.* Second edition. Interdisciplinary Applied Mathematics, 14. Springer-Verlag, New York.

31. Rodríguez, D. J. and L. Torres-Sorando (2001), Models of infectious diseases in spatially heterogenous environments. *Bull. Math. Biol.* **63**, 547–571.

32. Ruan, S. and W. Wang (2003), Dynamical behavior of an epidemic model with a nonlinear incidence rate. *J. Differential Equations* 188, 135–163.

33. Sattenspiel, L. and K. Dietz (1995), A structured epidemic model incorporating geographic mobility among regions. *Math. Biosci.* **128**, 71–91.

34. Sattenspiel, L. and D. Herring (2003), Simulating the effect of quarantine on the spread of the 1918–19 flu in central Canada. *Bull. Math. Biol.* **65**, 1–26.

35. Smith, H. L. and P. Waltman (1995), *The theory of the Chemostat,* Cambridge University Press.

36. Smith, H. L. (1995), *Monotone Dynamical Systems. An introduction to the theory of competitive and cooperative systems,* Math Surveys and Monographs 41, American Mathematical Society, Providence, RI.

37. Takeuchi, Y. (1986), Global stability in generalized Lotka-Volterra diffusion systems. *J. Math. Anal. Appl.* **116**, 209–221.

38. Takeuchi, Y., Wanbiao Ma and E. Beretta (2000), Global asymptotic properties of a delay *SIR* epidemic model with finite incubation times. *Nonlinear Anal.* **42**, 931–947.

39. Thieme, H. R. (1993), Persistence under relaxed point-dissipativity (with application to an endemic model), *SIAM J. Math. Anal.,* **24**, 407–435.

40. Van den Driessche, P. and J. Watmough (2000), A simple SIS epidemic model with a backward bifurcation. *J. Math. Biol.* **40**, 525–540.

41. Van den Driessche, P. and J. Watmough (2002), Reproduction numbers and sub-threshold endemic equilibria for compartmental models of disease transmission, *Math. Biosci.,* **180**, 29–48.

42. Wang, W. (2002a), Global behavior of an SEIRS epidemic model with time delays. *Appl. Math. Lett.* **15**, 423–428.

43. Wang, W. (2004), Population Dispersal and Disease Spread. *Discrete and Continuous Dynamical Systems Series B.* **4**, 797–804.

44. Wang, W. and Z. Ma (2002b), Global dynamics of an epidemic model with time delay. *Nonlinear Anal. Real World Appl.* **3**, 365–373.

45. Wang, W. and G. Mulone (2003), Threshold of disease transmission on a patch environment. *J. Math. Anal. Appl.* **285**, 321–335.

46. Wang, W. and S. Ruan (2004a), Simulating the SARS outbreak in Beijing with limited data. *J. theor. Biol.* **227**, 369–379.

47. Wang, W. and S. Ruan (2004b), Bifurcation in an epidemic model with constant removal rate of the infectives. *J. Math. Anal. Appl.* **291**, 775–793.
48. Wang, W. and X.-Q. Zhao (2004), An epidemic model in a patchy environment, *Math. Biosci.* **190**, 39–69.
49. Xiao, Y. and L. Chen (2001), Modeling and analysis of a predator-prey model with disease in the prey. Math. Biosci. **171**, 59–82.
50. Zhao, X.-Q. and Z.-J. Jing (1996), Global asymptotic behavior in some cooperative systems of functional differential equations, *Canadian Applied Mathematics Quarterly* **4**, 421–444.

5

Spatial-Temporal Dynamics in Nonlocal Epidemiological Models *

Shigui Ruan

5.1 Introduction

Throughout recorded history, nonindigenous vectors that arrive, establish, and spread in new areas have fomented epidemics of human diseases such as malaria, yellow fever, typhus, plague, and West Nile (Lounibos 2002). The spatial spread of newly introduced diseases is a subject of continuing interest to both theoreticians and empiricists. One strand of theoretical developments (e. g., Kendall 1965; Aronson and Weinberger 1975; Murray 1989) built on the pioneering work of Fisher (1937) and Kolmogorov et al. (1937) based on a logistic reaction-diffusion model to investigate the spread of an advantageous gene in a spatially extended population. With initial conditions corresponding to a spatially localized introduction, such models predict the eventual establishment of a well-defined invasion front which divides the invaded and uninvaded regions and moves into the uninvaded region with constant velocity.

Provided that very small populations grow in the same way or faster than larger ones, the velocity at which an epidemic front moves is set by the rate of divergence from the (unstable) disease-free state, and can thus be determined by linear methods (e. g., Murray 1989). These techniques have been refined by Diekmann (1978, 1979), Thieme (1977a, 1977b, 1979), van den Bosch et al. (1990), etc. who used a closely related renewal equation formalism to facilitate the inclusion of latent periods and more general and realistic transport models. Behind the epidemic front, most epidemic models settle to a spatially homogeneous equilibrium state in which all populations co-exist at finite abundances. In many cases, the passage from the epidemic front to co-existence passes through conditions where the local abundances of some or all of the players drops to truly microscopic levels. Local rekindling of the

---

* Research was partially supported by NSF grant DMS-0412047, NIH Grant P20-RR020770-01, the MITACS of Canada and a Small Grant Award at the University of Miami.

disease usually takes place not only because of the immigration of infectives but also by in situ infections produced by the non-biological remnants of previous populations.

Kermack and McKendrik (1927) proposed a simple deterministic model of a directly transmitted viral or bacterial agent in a closed population consisting of susceptibles, infectives, and recovereds. Their model leads to a nonlinear integral equation which has been studied extensively. The deterministic model of Barlett (1956) predicts a wave of infection moving out from the initial source of infection. Kendall (1957) generalized the Kermack–McKendrik model to a space-dependent integro-differential equation. Aronson (1977) argued that the three-component Kendall model can be reduced to a scalar one and extended the concept of asymptotic speed of propagation developed in Aronson and Weinberger (1975) to the scalar epidemic model. The Kendall model assumes that the infected individuals become immediately infectious and does not take into account the fact that most infectious diseases have an incubation period. Taking the incubation period into consideration, Diekmann (1978, 1979) and Thieme (1977a, 1977b, 1979) simultaneously proposed a nonlinear (double) integral equation model and extended Aronson and Weinberger's concept of asymptotic speed of propagation to such models. All these models are integral equations in which the spatial migration of the population or host was not explicitly modelled.

Spatial heterogeneities can be included by adding an immigration term where infective individuals enter the system at a constant rate. De Mottoni et al. (1979) and Busenberg and Travis (1983) considered a population in an open bounded region and assumed that the susceptible, infective, and removed individuals can migrate inside the region according to the rules of group migration. The existence of traveling waves in epidemic models described by reaction-diffusion systems has been extensively studied by many researchers, for example, Thieme (1980), Källen et al. (1985), Murray et al. (1986) and Murray and Seward (1992) studied the spatial spread of rabies in fox; Abramson et al. (2003) considered traveling waves of infection in the Hantavirus epidemics; Cruickshank et al. (1999), Djebali (2001), Hosono and Ilyas (1995) investigated the traveling waves in general SI epidemic models; Caraco et al. (2002) studied the spatial velocity of the epidemic of lyme disease; Greenfell et al. (2001) discussed the traveling waves in measles epidemics; etc.

In this article we try to provide a short survey on the spatial-temporal dynamics of nonlocal epidemiological models, include the classical Kermack–McKendrick model, the Kendall model given by differential and integral equations, the Diekmann–Thieme model described by a double integral equation, the diffusive integral equations proposed by De Mottoni et al. (1979) and Busenberg and Travis (1983), a vector-disease model described by a diffusive double integral equation (Ruan and Xiao 2004), etc.

5.2 Kermack–McKendrick model

Kermack and McKendrick (1927) proposed a simple deterministic model of a directly transmitted viral or bacterial agent in a closed population based on the following assumptions: (i) a single infection triggers an autonomous process within the host; (ii) the disease results in either complete immunity or death; (iii) contacts are according to the law of mass-action; (iv) all individuals are equally susceptible; (v) the population is closed in the sense that at the time-scale of disease transmission the inflow of new susceptibles into the population is negligible; (vi) the population size is large enough to warrant a deterministic description.

Let $S(t)$ denote the (spatial) density of individuals who are susceptible to a disease, that is, who are not yet infected at time t. Let $A(\theta)$ represent the expected infectivity of an individual who became infected θ time units ago. If $\frac{\mathrm{d}S}{\mathrm{d}t}(t)$ is the incidence at time t, then $\frac{\mathrm{d}S}{\mathrm{d}t}(t-\theta)$ is the number of individuals arising per time unit at time t who have been infected for θ time units. The original Kermack–McKendrick model is the following integral differential equation

$$\frac{\mathrm{d}S}{\mathrm{d}t} = S(t) \int_0^\infty A(\theta) \frac{\mathrm{d}S}{\mathrm{d}t}(t-\theta)\,\mathrm{d}\theta \,. \tag{1}$$

Kermack and McKendrick (1927) derived an invasion criterion based on the linearization of (1). Assume $S(0) = S_0$, the density of the population at the beginning of the epidemic with everyone susceptible. Suppose the solution of the linearized equation at S_0 has the form ce^{rt}. Then the characteristic equation is

$$1 = S_0 \int_0^\infty A(\theta)\,\mathrm{e}^{-r\theta}\,\mathrm{d}\theta \,.$$

Define

$$R_0 = S_0 \int_0^\infty A(\theta)\,\mathrm{d}\theta \,.$$

Here, R_0 is the number of secondary cases produced by one typical primary case and describes the growth of the epidemic in the initial phase on a generation phase. Since $A(\theta)$ is positive, we have $r > 0$ if and only if $R_0 > 1$. Therefore, the invasion criterion is $R_0 > 1$.

If the kernel $A(\theta)$ takes the special form $\beta e^{-\gamma\theta}$, where $\beta > 0, \gamma > 0$ are constants, and define

$$I(t) = -\frac{1}{\beta} \int_0^\infty A(\theta) \frac{\mathrm{d}S}{\mathrm{d}t}(t-\theta)\,\mathrm{d}\theta \,.$$

Then $I(t)$ represents the number of infected individuals at time t. Let $R(t)$ denote the number of individuals who have been infected and then removed from the possibility of being infected again or of spreading infection. Thus,

$\frac{\mathrm{d}R}{\mathrm{d}t} = \gamma I(t)$. This, together with (1) and differentiation of $I(t)$, yields the following ODE system

$$
\begin{aligned}
\frac{\mathrm{d}S}{\mathrm{d}t} &= -\beta S(t)I(t) \, , \\
\frac{\mathrm{d}I}{\mathrm{d}t} &= \beta S(t)I(t) - \gamma I(t) \, , \\
\frac{\mathrm{d}R}{\mathrm{d}t} &= \gamma I(t) \, .
\end{aligned}
\tag{2}
$$

Remark 1. Interestingly, it is system (2) (instead of the original equation (1)) that is widely referred to as the Kermack–McKendrick model. Though Kermack and McKendrick (1927) studied the special case (2), a more general version was indeed previously considered by Ross and Hudson (1917) (see Diekmann et al. 1995).

Observe from system (2) that $\frac{\mathrm{d}S}{\mathrm{d}t} < 0$ for all $t \geq 0$ and $\frac{\mathrm{d}I}{\mathrm{d}t} > 0$ if and only if $S(t) > \gamma/\beta$. Thus, $I(t)$ increases so long as $S(t) > \gamma/\beta$, but $S(t)$ decreases for all $t \geq 0$, it follows that $I(t)$ eventually decreases and approaches zero. Define the *basic reproduction number* as

$$
R_0 = \frac{\beta S(0)}{\gamma} \, .
$$

If $R_0 > 1$, then $I(t)$ first increases to a maximum attained when $S(t) = \gamma/\beta$ and then decreases to zero (epidemic). If $R_0 < 1$, then $I(t)$ decreases to zero (no epidemic).

The Kermack–McKendrick model and the threshold result derived from it have played a pivotal role in subsequent developments in the study of the transmission dynamics of infective diseases (Anderson and May 1991; Brauer and Castillo-Chavez 2000; Diekmann and Heesterbeek 2000; Hethcote 2000; Thieme 2003).

5.3 Kendall model

Kendall (1957, 1965) generalized the Kermack–McKendrick model to a space-dependent integro-differential equation. Denote $\mathbb{R} = (-\infty, \infty), \mathbb{R}_+ = [0, \infty)$. Let $S(x, t), I(x, t)$ and $R(x, t)$ denote the local densities of the susceptible, infected, and removed individuals at time t in the location $x \in \mathbb{R}$ with $S+I+R$ independent of t. All infected individuals are assumed to be infectious and the rate of infection is given by

$$
\beta \int_{-\infty}^{\infty} I(y, t)K(x - y)\,\mathrm{d}y \, ,
$$

where $\beta > 0$ is a constant and the kernel $K(x - y) > 0$ weights the contributions of the infected individuals at location y to the infection of susceptible individuals at location x. It is assumed that

$$\int_{-\infty}^{\infty} K(y)\,\mathrm{d}y = 1\,.$$

Removed individuals can be regarded as being either immune or dead and the rate of removal is assumed to be $\gamma I(x,t)$, where $\gamma > 0$ is a constant. With this notation, Kendall's model is

$$\frac{\partial S}{\partial t} = -\beta S(x,t) \int_{-\infty}^{\infty} I(y,t) K(x-y)\,\mathrm{d}y\,,$$

$$\frac{\partial I}{\partial t} = \beta S(x,t) \int_{-\infty}^{\infty} I(y,t) K(x-y)\,\mathrm{d}y - \gamma I(x,t)\,, \tag{3}$$

$$\frac{\partial R}{\partial t} = \gamma I(x,t)\,.$$

Given a spatially inhomogeneous epidemic model it is very natural to look for traveling wave solutions. The basic idea is that a spatially inhomogeneous epidemic model can give rise to a moving zone of transition from an infective state to a disease-free state. A *traveling wave solution* of system (3) takes the form $(S(x-ct,t), I(x-ct,t), R(x-ct,t))$. Kendall (1965) proved the existence of a positive number c^* such that the model admits traveling wave solutions of all speeds $c \geq c^*$ and no traveling wave solutions with speeds less than c^*. Mollison (1972) studied Kendall's original model in the special case in which there are no removals. With this assumption the system of integro-differential equations reduces to a single equation. For a particular choice of the averaging kernel Mollison (1972) proved the analog of Kendall's result. Atkinson and Reuter (1976) analyzed the full Kendall model for a general class of averaging kernels and obtained a criterion for the existence of a critical speed $c^* > 0$ and the existence of traveling waves of all speeds $c > c^*$. See also Barbour (1977), Brown and Carr (1977), Medlock and Kot (2003), etc.

Minimal wave speeds analogous to those found by Kendall and others also occur in the classical work of Fisher (1937) and Kolmogoroff et al. (1937) concerning the advance of advantageous genes. Aronson and Weinberger (1975, 1978) showed that the minimal wave speed is the *asymptotic speed* of propagation of disturbances from the steady state for Fisher's equation. Roughly speaking, $c^* > 0$ is called the asymptotic speed if for any c_1, c_2 with $0 < c_1 < c^* < c_2$, the solution tends to zero uniformly in the region $|x| \geq c_2 t$, whereas it is bounded away from zero uniformly in the region $|x| \leq c_1 t$ for t sufficiently large. Aronson (1977) proved that an analogous result holds for Kendall's epidemic model.

A steady state of system (3) is given by $S = \sigma, I = R = 0$, where $\sigma > 0$ is a constant. To study the asymptotic behavior of solutions to system (3), consider the initial values

$$S(x,0) = \sigma\,, \quad I(x,0) = I_0(x)\,, \quad R(x,0) = 0\,, \quad x \in \mathbb{R}\,, \tag{4}$$

where $I_0(x) \geq 0$ is continuous such that $I(x) \not\equiv 0$ and $I(x) \equiv 0$ in $[x_0, \infty)$ for some $x_0 \in \mathbb{R}$.

By rescaling, the initial value problem can be re-written as

$$\frac{\partial S}{\partial t} = -S(x,t) \int_{-\infty}^{\infty} I(y,t) K(x-y) \, dy \,,$$

$$\frac{\partial I}{\partial t} = S(x,t) \int_{-\infty}^{\infty} I(y,t) K(x-y) \, dy - \lambda I(x,t) \,,$$

$$\frac{\partial R}{\partial t} = \lambda I(x,t) \,,$$

$$S(x,0) = 1 \,, \quad I(x,0) = I_0(x) \,, \quad R(x,0) = 0 \,, \quad x \in \mathbb{R} \,,$$

$$(5)$$

where $\lambda = \gamma/\beta\sigma$. It is not difficult to see that if (S, I, R) is a solution of system (5), then R satisfies

$$\frac{\partial R}{\partial t} = -\lambda R(x,t) + \lambda \left[1 - \exp\left(-\frac{1}{\lambda} \int_{-\infty}^{\infty} R(y,t) K(x-y) \, dy \right) \right] + \lambda I_0(x) \,,$$

$$R(x,0) = 0 \,, \quad x \in \mathbb{R} \,.$$

$$(6)$$

Conversely, if R is a solution of the problem (6), then (S, I, R) is a solution of system (5) with

$$S = \exp\left(-\frac{1}{\lambda} \int_{-\infty}^{\infty} R(y,t) K(x-y) \, dy \right) \,,$$

$$I = -R + \left[1 - \exp\left(-\frac{1}{\lambda} \int_{-\infty}^{\infty} R(y,t) K(x-y) \, dy \right) \right] + I_0(x) \,.$$

Assume that

(K1) K is a nonnegative even function defined in $\mathbb{R}$ with $\int_{-\infty}^{\infty} K(y) \, dy = 1$.

(K2)   There exists a $\nu \in (0, \infty]$ such that $\int_{-\infty}^{\infty} e^{\mu y} K(y) \, dy < \infty$ for all $\mu \in [0, \nu)$.

(K3)   Define $A_\lambda(\mu) = \frac{1}{\mu}[\int_{-\infty}^{\infty} e^{\mu y} K(y) \, dy - \lambda]$. For each $\lambda < 1$ there exists a $\mu^* = \mu^*(\lambda) \in (0, \nu)$ such that $0 < c^* \equiv A_\lambda(\mu^*) = \inf\{A_\lambda(\mu) : 0 < \mu < \nu\}$, $A'_\lambda(\mu) < 0$ in $(0, \mu^*)$ and $A'_\lambda(\mu) > 0$ in (μ^*, ν).

(K4)   For each $\bar{\mu} \in (0, \nu)$ there exists an $r = r(\bar{\mu}) \geq 0$ such that $e^{\mu x} K(x) = \min\{e^{\mu y} K(y) : y \in [0, x]\}$ for all $\mu \in [0, \bar{\mu}]$ and $x \geq r(\bar{\mu})$.

Theorem 1 (Aronson 1977). *Suppose the kernel K satisfies (K1)–(K4). Let $R(x,t)$ be a solution of the problem (6). If $\lambda \geq 1$, then for every $x \in \mathbb{R}$ and $c \geq 0$,*

$$\lim_{t \to \infty, |x| \geq ct} R(x,t) = 0 \,.$$

Theorem 1 corresponds to the Kermack–McKendrick threshold result. Roughly speaking, it says that an initial infection (given by $I_0(x)$) does not propagate if $\lambda = \gamma/\beta\sigma \geq 1$, that is, if the initial density of susceptibles (σ) is too low or the removal rate (γ) is too high.

The next result shows that the situation is quite different for $\lambda \in (0, 1)$.

Theorem 2 (Aronson 1977). *Suppose the kernel K satisfies (K1)–(K4) and $\lambda \in (0, 1)$. Let $R(x, t)$ be a solution of the problem (6).*

(i) If $c > c^*$, then for every $x \in \mathbb{R}$,

$$\lim_{t \to \infty, |x| \geq ct} R(x, t) = 0\,.$$

(ii) If $0 < c < c^*$, then for every $x \in \mathbb{R}$,

$$\lim_{t \to \infty, |x| \geq ct} R(x, t) = \alpha(\lambda)\,,$$

where $\alpha(\lambda)$ is the unique solution of $1 - \alpha = e^{-\alpha/\lambda}$ in $(0, 1)$.

Theorem 2 says that if you travel toward $+\infty$ from any point in $\mathbb{R}$, then you will outrun the infection if your speed exceeds the minimal speed c^*, but the infection will overtake you if your speed is less than c^*.

5.4 Diekmann–Thieme model

Suppose that not all individuals are equally susceptible, but certain traits have a marked influence. Let $S(x, t)$ denote the density of susceptibles at time t and location x and $i(x, t, \theta)\,d\tau$ be the density of infectives who were infected some time between $t - \theta$ and $t - \theta - d\theta$. Then $I(x, t) = \int_0^\infty i(x, t, \theta)\,d\theta$ is the density of infectives at time t and location x. Let $A(\theta, x, y)$ represent the expected infectivity of an individual who became infected θ time units ago while having a trait value y towards a susceptible with trait value x. Similar to the Kermack–McKendrick model (1), one has (Diekmannn 1978)

$$\frac{\partial S}{\partial t} = S(x, t) \int_\Omega \int_0^\infty A(\theta, x, y) \frac{\partial S}{\partial t}(y, t - \theta)\,d\theta\,dy\,, \tag{7}$$

where Ω denotes the set of trait values. Assume

$$i(x, 0, \theta) = i_0(x, \theta)\,, \quad S(x, 0) = S_0(x)\,.$$

Then (7) can be written as

$$\frac{\partial S}{\partial t} = S(x, t) \left[\int_0^t \int_\Omega A(\theta, x, y) \frac{\partial S}{\partial t}(y, t - \theta)\,dy\,d\theta - h(x, t) \right]\,, \tag{8}$$

where

$$h(x,t) = \int_0^\infty \int_\Omega i_0(x,\theta) A(t+\theta,x,y)\,\mathrm{d}y\,\mathrm{d}\theta\,.$$

Now, assuming $S_0(x) > 0$ for every $x \in \Omega$ and integrating equation (8) with respect to t, one obtains the Diekmann–Thieme model (Diekmann 1978, 1979; Thieme 1977a, 1977b, 1979)

$$u(x,t) = \int_0^t \int_\Omega g(u(y,t-\theta))k(\theta,x,y)\,\mathrm{d}y\,\mathrm{d}\theta + f(x,t)\,, \qquad (9)$$

where

$$u(x,t) = -\ln\frac{S(x,t)}{S_0(x)}\,, \qquad g(u) = 1 - \mathrm{e}^{-u}\,,$$

$$k(\theta,x,y) = S_0(y)A(\theta,x,y)\,, \quad f(x,t) = \int_0^t h(x,s)\,\mathrm{d}s\,.$$

Let $BC(\Omega)$ be the Banach space of bounded continuous functions on Ω equipped with the supremum norm. Denote $C_T = C([0,T]; BC(\Omega))$ the Banach space of continuous functions on $[0,T]$ with values in $BC(\Omega)$ equipped with the norm

$$\|f\|_{C_T} = \sup_{0\le t\le T} \|f[t]\|_{BC(\Omega)}\,,$$

where $f(x,t)$ is written as $f[t]$ when it is regarded as an element of C_T. The first result is about the local and global existence and uniqueness of the solution of (9).

Theorem 3 (Diekmann 1978). *Suppose g is locally Lipschitz continuous and $f\colon \mathbb{R}_+ \to BC(\Omega)$ is continuous, then there exists a $T > 0$ such that (9) has a unique solution u in C_T. If g is uniformly Lipschitz continuous, then (9) has a unique solution $u\colon \mathbb{R}_+ \to BC(\Omega)$.*

Remark 2. Thieme (1977a) proved a very similar result for a more general model and considered how far an epidemic can spread. See also Thieme (1977b).

The next result is about the positivity, monotonicity and stabilization of the solution of equation (9).

Theorem 4 (Diekmann 1978).

(1) Suppose $g(u) > 0$ for $u > 0$ and $f[t] \ge 0$ for all $t \ge 0$, then $u[t] \ge 0$ on the domain of definition of u.

(2) Suppose, in addition, g is monotone nondecreasing and $f[t+h] \ge f[t]$ for all $h \ge 0$, then $u[t+h] \ge u[t]$ for all $h \ge 0$ and $t \ge 0$ such that $t+h$ is in the domain of definition of u.

(3) Suppose, in addition, that g is bounded and uniformly Lipschitz continuous on $\mathbb{R}_+$ and that the subset $\{f[t]|t \geq 0\}$ of $BC(\Omega)$ is uniformly bounded and equicontinuous and that k satisfies

(i) For each $x \in \Omega$, $\int_0^t k(\theta, x, \cdot)\,\mathrm{d}\theta \to \int_0^\infty k(\theta, x, \cdot)\,\mathrm{d}\theta$ in $L_1(\Omega)$ as $t \to \infty$, and for some $C > 0$, $\sup_{x \in \Omega} \int_\Omega \int_0^\infty k(\theta, x, y)\,\mathrm{d}\theta\,\mathrm{d}y < C$.

(ii) For each $\varepsilon > 0$ there exists $\delta = \delta(\varepsilon) > 0$ such that if $x_1, x_2 \in \Omega$ and $|x_1 - x_2| < \delta$, then $\int_\Omega \int_0^\infty |k(\theta, x_1, y) - k(\theta, x_2, y)|\,\mathrm{d}\theta\,\mathrm{d}y < \varepsilon$.

Then the solution u of (9) is defined on $\mathbb{R}_+$ and there exists $u[\infty] \in BC(\Omega)$ such that, as $t \to \infty$, $u[t] \to u[\infty]$ in $BC(\Omega)$ if Ω is compact, and uniformly on compact subset of Ω if Ω is not compact. Moreover, $u[\infty]$ satisfies the limit equation

$$u[\infty] = \int_\Omega g(u[\infty](y)) \int_0^\infty k(\theta, x, y)\,\mathrm{d}\theta\,\mathrm{d}y + f[\infty](x)\,.$$

Now consider the Diekmann–Thieme model (9) with $\Omega = \mathbb{R}^n (n = 1, 2, 3)$.

Assume $k(\theta, x, y) = k(\theta, x - y) \colon \mathbb{R}_+ \times \mathbb{R}^n \to \mathbb{R}_+$ is a Borel measurable function satisfying

(k1) $k^* = \int_0^\infty \int_{\mathbb{R}^n} k(\theta, y)\,\mathrm{d}y\,\mathrm{d}\theta \in (1, \infty)$.

(k2) There exists some $\lambda_0 > 0$ such that $\int_0^\infty \int_{\mathbb{R}^n} e^{\lambda_0 y_1} k(\theta, y)\,\mathrm{d}y\,\mathrm{d}\theta < \infty$, where y_1 is the first coordinate of $y \in \mathbb{R}^n$.

(k3) There are constants $\sigma_2 > \sigma_1 > 0, \rho > 0$ such that $k(\theta, x) > 0$ for all $\theta \in (\sigma_1, \sigma_2)$ and $|x| \in (0, \rho)$.

(k4) k is isotropic (i.e., $k(\theta, x) = k(\theta, y)$ if $|x| = |y|$).

Define

$$c^* = \inf \left\{ c \geq 0 : \int_0^\infty \int_{\mathbb{R}^n} e^{-\lambda(c\theta + y_1)} k(\theta, y)\,\mathrm{d}y\,\mathrm{d}\theta < 1 \text{ for some } \lambda > 0 \right\}. \quad (10)$$

Assume that $g \colon \mathbb{R}_+ \to \mathbb{R}_+$ is a Lipschitz continuous function satisfying

(g1) $g(0) = 0$ and $g(u) > 0$ for all $u > 0$.

(g2) g is differentiable at $u = 0, g'(0) = 1$ and $g(u) \leq u$ for all $u > 0$.

(g3) $\lim_{u \to \infty} g(u)/u = 0$.

(g4) There exists a positive solution u^* of $u = k^* g(u)$ such that $k^* g(u) > u$ for all $u \in (0, u^*)$ and $k^* g(u) < u$ for all $u > u^*$.

Thieme (1979) proved that the c^* defined by (10) is the asymptotic wave speed (see also Diekmann 1979; Thieme and Zhao 2003).

Theorem 5 (Thieme 1979). *Assume k satisfies (k1)–(k4) and g satisfies (g1)–(g4).*

(i) For every admissible $f(x, t)$, the unique solution $u(x, t)$ of (9) satisfies

$$\lim_{t \to \infty, |x| \geq ct} u(x, t) = 0$$

for each $c > c^*$.

(ii) If g is monotone increasing and $f(x,t)\colon \mathbb{R}^n \times \mathbb{R}_+ \to \mathbb{R}_+$ is a Borel measurable function such that $f(x,t) \geq \eta > 0$ for all $t \in (t_1, t_2)$ and $|x| \leq \eta$ with $t_2 > t_1 \geq 0, \eta > 0$, then

$$\lim_{t \to \infty, |x| \geq ct} u(x,t) \geq u^*$$

for each $c \in (0, c^*)$.

To discuss the existence of traveling wave solutions in equation (9), we assume $\Omega = \mathbb{R}$ and $f(x,t) = 0$. Suppose g satisfies the modified assumptions:

(g5) $g(0) = 0$ and there exists a positive solution u^* of $u = k^* g(u)$ such that $k^* g(u) > u$ for all $u \in (0, u^*)$.

(g6) g is differentiable at $u = 0, g'(0) = 1$ and $g(u) \leq u$ for all $u \in [0, u^*]$.

Theorem 6 (Diekmann 1978, 1979). *Suppose $k(\theta, x)$ satisfies (k1)-(k4) with $n = 1$ and g satisfies (g5)-(g6). Moreover, assume that g is monotone increasing on $[0, u^*]$ and $g(u) \geq u - a\, u^2$ for all $u \in [0, u^*]$ and some $a > 0$. Then for each $c > c^*$, there exists a montone traveling wave solution of (9) with speed c which connects 0 and u^*.*

Remark 3. Thieme and Zhao (2003) considered a more general nonlinear integral equation and studied the asymptotic speeds of spread and traveling waves. Schumacher (1980a, 1980b) argued that the following model

$$\frac{\partial u}{\partial t} = \int_0^\infty \int_{-\infty}^\infty g(u(x - y, t - s))\, \mathrm{d}\eta\,(y, s) \tag{11}$$

is more reasonable, where η is a Lebesque measure on $\mathbb{R} \times \mathbb{R}_+$ such that $\eta(\mathbb{R} \times \mathbb{R}_+) = 1$, and investigated the asymptotic speed of propagation, existence of traveling fronts and dependence of the minimal speed on delays.

5.5 Migration and spatial spread

Spatial heterogeneities can be included by adding an immigration term where infective individuals enter the system at a constant rate. This clearly allows the persistence of the disease because if it dies out in one region then the arrival of an infective from elsewhere can trigger another epidemic. Indeed, the arrival of new infectives has been demonstrated as being important in the outbreaks of measles observed in Iceland, a small island community (Cliff et al., 1993). A constant immigration term has a mildly stabilizing effect on the dynamics and tends to increase the minimum number of infective individuals observed in the models (Bolker and Grenfell 1995). De Mottoni et al. (1979) and Busenberg and Travis (1983) considered a population in an open bounded region $\Omega \subset \mathbb{R}^n (n \leq 3)$ with smooth boundary $\partial\Omega$ and assumed that the susceptible, infective, and removed individuals can migrate inside the region Ω according to the rules of group migration.

5.5.1 An SI model

Assume the population consists of only two classes, the susceptibles $S(x,t)$ and the infectives $I(x,t)$, at time t and location $x \in \Omega$. Assume that both the susceptibles and infectives can migrate according to a Fickian diffusion law with each subpopulation undergoing a flux which is proportional to the gradient of that particular subpopulation: ΔS and $d\Delta I$, respectively, where the diffusion rate of the susceptibles is normalized to be one and $d > 0$ is the diffusion rate for the infectives. The mechanism of infection is governed by a nonlocal law, as in the Kendall model. It is also assumed that the susceptibles grow at a rate $\mu > 0$ the susceptibles are removed (e.g. by vaccination) depending on an effectiveness coefficient σ. Based on these assumptions, De Mottoni et al. (1979) considered the following model

$$\frac{\partial S}{\partial t} = \Delta S + \mu - \sigma S(x,t) - S(x,t)\int_{\Omega} I(y,t)K(x,y)\,\mathrm{d}y \,,$$

$$\frac{\partial I}{\partial t} = \mathrm{d}\Delta I + S(x,t)\int_{\Omega} I(y,t)K(x,y)\,\mathrm{d}y - \gamma I(x,t) \tag{12}$$

under the boundary value conditions

$$\frac{\partial S}{\partial n}(x,t) = \frac{\partial I}{\partial n}(x,t) = 0 \,, \quad (x,t) \in \partial\Omega \times (0,\infty) \tag{13}$$

and initial value conditions

$$S(x,0) = S_0(x) \,, \quad I(x,0) = I_0(x) \,, \quad x \in \Omega \,. \tag{14}$$

Let $C(\bar{\Omega})$ denote the Banach space of continuous functions on $\bar{\Omega}$ endowed with supremum norm $\|u\| = \max_{x\in\bar{\Omega}} |u(x)|$. Let $X = C(\bar{\Omega}) \oplus C(\bar{\Omega})$ with norm $|U|_X = \|u\| + \|v\|$ for $U = (u,v) \in X$.

De Mottoni et al. (1979) proved the following local stability and global attractivity of the disease free equilibrium $(\mu/\sigma, 0)$, where stability is meant relative to the X-norm. Thus, the threshold type result has been generalized to the diffusive nonlocal epidemic model (12).

Theorem 7 (De Mottoni et al. 1979). *Assume that $\mu < \gamma/\sigma$. Then*

(i)   The steady state solution $(\mu/\sigma, 0)$ is asymptotically stable.

(ii) For any $(S_0, I_0) \in X$ with $S_0 \geq 0, I_0 \geq 0$, the corresponding solution of (12) converges to $(\mu/\sigma, 0)$ in X as $t \to \infty$.

When $\mu = \sigma = 0, K(\cdot)$ equals β times a delta function, system (12) reduces to a reaction-diffusion model of the form

$$\frac{\partial S}{\partial t} = \frac{\partial^2 S}{\partial x^2} - \beta SI \,,$$

$$\frac{\partial I}{\partial t} = d\frac{\partial^2 I}{\partial x^2} + \beta SI - \gamma I \,. \tag{15}$$

Capasso (1979) and Webb (1981) studied the stability of the disease free steady state of the system (15).

To discuss the existence of traveling wave solutions, consider $x \in \mathbb{R}$. Look for traveling wave solutions of the form

$$S(x,t) = g(\xi), \quad I(x,t) = f(\xi), \quad \xi = x - ct$$

satisfying

$$g(-\infty) = \varepsilon(\varepsilon < S_0), \quad g(+\infty) = S_0, \quad f(-\infty) = f(+\infty) = 0. \tag{16}$$

where c is the wave speed to be determined, ε is some positive constant. The following result was obtained by Hosono and Ilyas (1995).

Theorem 8 (Hosono and Ilyas 1995). *Assume that $\gamma/\beta S_0 < 1$. Then for each $c \geq c^* = 2\sqrt{\beta S_0 \, d(1 - \gamma/\beta S_0)}$ there exists a positive constant ε^* such that system (15) has a traveling wave solution $(S(x,t), I(x,t)) = (g(\xi), f(\xi))$ for $\varepsilon = \varepsilon^*$.*

Notice that when $\gamma/\beta S_0 > 1$, the system has no traveling wave solutions. The threshold condition $\gamma/\beta S_0 < 1$ for the existence of traveling wave solutions has some implications. We can see that for any epidemic wave to occur, there is a minimum critical density of the susceptible population $S_c = \gamma/\beta$. Also, for a given population size S_0 and mortality rate γ, there is a critical transmission rate $\beta_c = \gamma/S_0$. When $\beta > \beta_c$, the infection will spread. With a given transmission rate and susceptible population we can also obtain a critical mortality rate $\gamma_c = \beta S_0$, there is an epidemic wave moving through the population if $\gamma < \gamma_c$.

5.5.2 An SIR model

Assume that a portion of those who are infected acquire immunity to further infection and join the removed class, while the remainder of those who are infected return to the susceptible class and are subject to possible further infections. Busenberg and Travis (1983) derived the following Kendall type SIS model

$$\frac{\partial S}{\partial t} = d\frac{S(x,t)}{N(x,t)}\Delta S - S(x,t)\int_\Omega I(y,t)K(x,y)\,\mathrm{d}y + \gamma_1 I(x,t),$$

$$\frac{\partial I}{\partial t} = d\frac{I(x,t)}{N(x,t)}\Delta I + S(x,t)\int_\Omega I(y,t)K(x,y)\,\mathrm{d}y - \gamma I(x,t),$$

$$\frac{\partial R}{\partial t} = d\frac{R(x,t)}{N(x,t)}\Delta I + \gamma_2 I(x,t) \tag{17}$$

under the boundary value conditions

$$\frac{\partial S}{\partial n}(x,t) = \frac{\partial I}{\partial n}(x,t) = \frac{\partial R}{\partial n}(x,t) = 0, \quad (x,t) \in \partial\Omega \times (0,\infty) \tag{18}$$

and initial value conditions

$$S(x,0) = S_0(x)\,, \quad I(x,0) = I_0(x)\,, \quad R(x,0) = R_0(x)\,, \quad x \in \Omega\,. \quad (19)$$

$N(x,t) = S(x,t) + I(x,t) + R(x,t)$ satisfies the linear initial-boundary value problem

$$\frac{\partial N}{\partial t} = d\Delta N(x,t)\,,$$

$$\frac{\partial N}{\partial n}(x,t) = 0\,, \quad (x,t) \in \partial\Omega \times (0,\infty)\,, \quad (20)$$

$$N(x,0) = S_0(x) + I_0(x) + R_0(x)\,, \quad x \in \Omega\,.$$

Theorem 9 (Busenberg and Travis 1983). *Let $K(x,y) > 0$ be twice continuously differentiable on $\bar\Omega \times \bar\Omega$, and let $S_0 > 0, I_0 > 0, R_0 > 0$ be twice continuously differentiable with the sum N_0 satisfying the Nuemann condition in (20). Then the problem (17)–(19) has a unique positive solution $(S(x,t), I(x,t), R(x,t))$ for $(x,t) \in \Omega \times \mathbb{R}_+$. Moreover,*

$$\lim_{t\to\infty} (S(x,t), I(x,t), R(x,t)) = (\hat{S}(x), \hat{I}(x), \hat{R}(x))\,, \quad (21)$$

where

$$\hat{S}(x) = \frac{a_0(N_0(x) - R_0(x))}{N_0(x)} - a_0\gamma_2 \int_0^\infty \frac{I(x,s)}{N(x,s)}\,\mathrm{d}s\,,$$

$$\hat{I}(x) = 0\,,$$

$$\hat{R}(x) = \frac{a_0 R_0(x)}{N_0(x)} + a_0\gamma_2 \int_0^\infty \frac{I(x,s)}{N(x,s)}\,\mathrm{d}s$$

and $a_0 = \int_\Omega N_0(x)\,\mathrm{d}x / \int_\Omega \mathrm{d}x$.

The result indicates that a portion of those who are infected eventually acquire immunnity, and the only possible limit is one where the disease dies out. The steady state distribution of the susceptible and immune subpopulations is generally spatially non-uniform and depends on the initial distributions of the different subclasses. It also depends on the time history of the evolution of the proportion $I(x,t)/N(x,t)$ of infected individuals through the integral $a_0\gamma_2 \int_0^\infty I(x,s)/N(x,s)ds$, which represents that portion of the infected subpopulation at position $x \in \Omega$ which becomes immune during the span of the epidemic.

5.6 A vector-disease model

We consider a host-vector model for a disease without immunity in which the current density of infectious vectors is related to the number of infectious hosts at earlier times. Spatial spread in a region is modeled by a diffusion term. Consider a host in a bounded region $\Omega \subset \mathbb{R}^n (n \leq 3)$ where a disease (malaria) is carried by a vector (mosquito). The host is divided into two classes, susceptible and infectious, whereas the vector population is divided into three classes, infectious, exposed, and susceptible. Suppose that the infection in the host confers negligible immunity and does not result death or isolation. All new-borns are susceptible. The host population is assumed to be stable, that is, the birth rate is constant and equal to the death rate. Moreover, the total host population is homogeneously distributed in Ω and both susceptible and infectious populations are allowed to diffuse inside Ω, however, there is no migration through $\partial\Omega$, the boundary of Ω.

For the transmission of the disease, it is assumed that a susceptible host can receive the infection only by contacting with infected vectors, and a susceptible vector can receive the infection only from the infectious host. Also, a susceptible vector becomes exposed when it receives the infection from an infected host. It remains exposed for some time and then becomes infectious. The total vector population is also constant and homogeneous in Ω. All three vector classes diffuse inside Ω and cannot cross the boundary of Ω.

Denote by $u(t,x)$ and $v(t,x)$ the normalized spatial density of infectious and susceptible host at time t in x, respectively, where the normalization is done with respect to the spatial density of the total population. Hence, we have

$$u(t,x) + v(t,x) = 1 , \quad (t,x) \in \mathbb{R}_+ \times \Omega .$$

Similarly, define $I(t,x)$ and $S(t,x)$ as the normalized spatial density of infectious and susceptible vector at time t in x, respectively.

If α denotes the host-vector contact rate, then the density of new infections in host is given by

$$\alpha v(t,x)I(t,x) = \alpha[1 - u(t,x)]I(t,x) .$$

The density of infections vanishes at a rate $au(t,x)$, where a is the cure/recovery rate of the infected host. The difference of host densities of arriving and leaving infections per unit time is given by $d\Delta u(t,x)$, where d is the diffusion constant, Δ is the Laplacian operator. We then obtain the following equation

$$\frac{\partial u}{\partial t}(t,x) = d\Delta u(t,x) - au(t,x) + \alpha[1 - u(t,x)]I(t,x) . \tag{22}$$

If the vector population is large enough, we can assume that the density of vectors which become exposed at time t in $x \in \Omega$ is proportional to the density of the infectious hosts at time t in x. That is, $E(t,x) = hu(t,x)$, where h is a positive constant. Let $\xi(t,s,x,y)$ denote the proportion of vectors which arrive in x at time t, starting from y at time $t-s$, then

$$\int_\Omega \xi(t,s,x,y)E(t-s,y)\,\mathrm{d}y$$

is the density of vectors which became exposed at time $t - s$ and are in x at time t. Let $\eta(s)$ be the proportion of vectors which are still infectious s units of time after they became exposed, then

$$
\begin{aligned}
I(t, x) &= \int_0^\infty \int_\Omega \xi(t, s, x, y) E(t - s, y) \eta(s) \, dy \, ds \\
&= \int_0^\infty \int_\Omega \xi(t, s, x, y) h \eta(s) u(t - s, y) \, dy \, ds \, .
\end{aligned}
$$

Substituting $I(t, x)$ into (22), changing the limits, and denoting

$$
b = \alpha h \, , \quad F(t, s, x, y) = \xi(t, s, x, y) \eta(s) \, ,
$$

we obtain the following diffusive integro-differential equation modeling the vector disease

$$
\frac{\partial u}{\partial t}(t, x) = d\Delta u(t, x) - a u(t, x) + b[1 - u(t, x)] \int_{-\infty}^t \int_\Omega F(t, s, x, y) u(s, y) \, dy \, ds
$$

$$\tag{23}$$

for $(t, x) \in I\!R_+ \times \Omega$. The initial value condition is given by

$$
u(0, x) = \phi(0, x) \, , \quad (\theta, x) \in (-\infty, 0] \times \Omega \, , \tag{24}
$$

where ϕ is a continuous function for $(\theta, x) \in (-\infty, 0] \times \Omega$, and the boundary value condition is given by

$$
\frac{\partial u}{\partial n}(t, x) = 0 \, , \quad (t, x) \in \mathbb{R}_+ \times \partial\Omega \, , \tag{25}
$$

where $\partial/\partial n$ represents the outward normal derivative on $\partial\Omega$.

The convolution kernel $F(t, s, x, y)$ is a positive continuous function in its variables $t \in \mathbb{R}, s \in \mathbb{R}_+, x, y \in \Omega$. We normalize the kernel so that

$$
\int_0^\infty \int_\Omega F(t, s, x, y) \, dy \, ds = 1 \, .
$$

Various types of equations can be derived from (23) by taking different kernels.

(i) If $F(t, s, x, y) = \delta(x - y) G(t, s)$, then (23) becomes the following integro-differential equation with a local delay

$$
\frac{\partial u}{\partial t} = d\Delta u(t, x) - a u(t, x) + b[1 - u(t, x)] \int_{-\infty}^t G(t - s) u(s, x) \, ds \tag{26}
$$

for $(t, x) \in \mathbb{R}_+ \times \Omega$.

(ii) If $F(t, s, x, y) = \delta(x - y) \delta(t - s)$, then (23) becomes the following reaction diffusion equation without delay

$$
\frac{\partial u}{\partial t} = d\Delta u(t, x) - a u(t, x) + b[1 - u(t, x)] u(t, x) \, , \quad (t, x) \in \mathbb{R}_+ \times \Omega \, . \tag{27}
$$

(iii) If $F(t, s, x, y) = \delta(x - y)\delta(t - s - \tau)$, where $\tau > 0$ is a constant, and u does not depend on the spatial variable, then (23) becomes the following ordinary differential equation with a constant delay

$$\frac{du}{dt} = -au(t) + b[1 - u(t)]u(t - \tau) . \tag{28}$$

Cooke (1977) studied the stability of (28) and showed that when $0 < b \leq a$, the trivial equilibrium $u_0 = 0$ is globally stable; when $0 \leq a < b$, the trivial equilibrium is unstable and the positive equilibrium $u_1 = (b - a)/b$ is globally stable. Busenberg and Cooke (1978) assumed that the coefficients are periodic and investigated the existence and stability of periodic solutions the (28). Thieme (1988) considered (28) when the coefficients are time-dependent and showed that, under suitable assumptions, the following dichotomy holds: either all non-negative solutions converge to zero or all pairs of non-negative solutions $u(t)$ and $v(t)$ with non-zero initial data satisfy $u(t)/v(t) \to 0$ as $t \to \infty$. The case with multiple groups and distributed risk of infection was studied by Thieme (1985). Marcati and Pozio (1983) proved the global stability of the constant solution to (23) when the delay is finite. Volz (1982) assumed that all coefficients are periodic and discussed the existence and stability of periodic solutions of (23).

We first consider the stability of steady states of (23) with a general kernel. Then we discuss the existence of traveling wave solutions in the equation when the kernel takes some specific forms.

5.6.1 Stability of the steady states

Denote $E = C(\bar{\Omega}, \mathbb{R})$. Then E is a Banach space with respect to the norm

$$|u|_E = \max_{x \in \bar{\Omega}} |u(x)| , \quad u \in E .$$

Denote $\mathcal{C} = BC((-\infty, 0], E)$. For $\phi \in \mathcal{C}$, define

$$\|\phi\| = \sup_{\theta \in (-\infty, 0]} |\phi(\theta)|_E .$$

For any $\beta \in (0, \infty)$, if $u : (-\infty, \beta) \to E$ is a continuous function, u_t is defined by $u_t(\theta) = u(t + \theta) , \quad \theta \in (-\infty, 0]$.

Define

$$\mathcal{D}(A) = \{u \in E : \Delta u \in E , \frac{\partial u}{\partial n} = 0 \text{ on } \partial\Omega\} ,$$

$$Au = d\Delta u \quad \text{for all} \quad u \in \mathcal{D}(A) ,$$

$$f(\phi)(x) = -a\phi(0, x) + b[1 - \phi(0, x)] \int_{-\infty}^{0} \int_{\Omega} F(0, s, x, y)\phi(s, y) \, dy \, ds ,$$

where $\phi \in C, x \in \bar{\Omega}$. Then we can re-write (23) into the following abstract form:

$$\frac{du}{dt} = Au + f(u_t), \quad t \geq 0,$$

$$u_0 = \phi \in C,$$

(29)

where

(a) $A \colon \mathcal{D}(A) \to E$ is the infinitesimal generator of a strongly continuous semigroup e^{tA} for $t \geq 0$ on E endowed with the maximum norm;
(b) $f \colon C \to E$ is Lipschitz continuous on bounded sets of C.

Associated to (29), we also consider the following integral equation

$$u(t) = e^{tA}\phi(0) + \int_0^t e^{(t-s)A} f(u_s)\,\mathrm{d}s, \quad t \geq 0,$$

$$u_0 = \phi.$$

(30)

A continuous solution of the integral equation (30) is called a *mild solution* to the abstract equation (29). The existence and uniqueness of the maximal mild solution to (29) follow from a standard argument (see Ruan and Wu (1994) and Wu (1996)). When the initial value is taken inside an invariant bounded set in C, the boundedness of the maximal mild solution implies the global existence.

Define

$$M = \left\{u \in E \colon 0 \leq u(x) \leq 1, x \in \bar{\Omega}\right\}.$$

We can prove that M is invariant by using the results on invariance and attractivity of sets for general partial functional differential equations established by Pozio (1980, 1983) and follow the arguments in Marcati and Pozio (1980).

Theorem 10 (Ruan and Xiao 2004). *The set M is invariant; that is, if $\phi \in BC((-\infty, 0]; M)$ then $u(\phi)$ exists globally and $u(\phi)(t) \in M$ for all $t \geq 0$.*

The stability of the steady state solutions can be established following the attractivity results of Pozio (1980, 1983).

Theorem 11 (Ruan and Xiao 2004). *The following statements hold*

(i)   *If $0 < b \leq a$, then $u_0 = 0$ is the unique steady state solution of (23) in M and it is globally asymptotically stable in $BC((-\infty, 0]; M)$.*
(ii) *If $0 \leq a < b$, then there are two steady state solutions in M : $u_0 = 0$ and $u_1 = (b-a)/b$, where u_0 is unstable and u_1 is globally asymptotically stable in $BC((-\infty, 0]; M)$.*

Recall that b represents the contact rate and a represents the recovery rate. The stability results indicate that there is a *threshold* at $b = a$. If $b \leq a$, then the proportion u of infectious individuals tends to zero as t becomes large and the disease dies out. If $b > a$, the proportion of infectious individuals tends to an endemic level $u_1 = (b - a)/b$ as t becomes large. There is no non-constant periodic solutions in the region $0 \leq u \leq 1$.

The above results also apply to the special cases (26), (27), and (28) and thus include the following results on global stability of the steady states of the discrete delay model (28) obtained by Cooke (1977) (using the Liapunov functional method).

Corollary 1 (Cooke 1977). *For the discrete delay model (28), we have the following statements*

(i) If $0 < b \leq a$, then the steady state solution $u_0 = 0$ is asymptotically stable and the set $\{\phi \in C([-\tau, 0], \mathbb{R}) : 0 \leq \phi(\theta) \leq 1 \text{ for } -\tau \leq \theta \leq 0\}$ is a region of attraction.

(ii) If $0 \leq a < b$, then the steady state solution $u_1 = (b-a)/b$ is asymptotically stable and the set $\{\phi \in C([-\tau, 0], \mathbb{R}) : 0 < \phi(\theta) \leq 1 \text{ for } -\tau \leq \theta \leq 0\}$ is a region of attraction.

5.6.2 Existence of traveling waves

We know that when $b > a$ (23) has two steady state solutions, $u_0 = 0$ and $u_1 = (b - a)/b$. In this section we consider $x \in (-\infty, \infty)$ and establish the existence of traveling wave solutions of the form $u(x,t) = U(z)$ such that

$$\lim_{z \to -\infty} U(z) = \frac{b - a}{b}, \quad \lim_{z \to \infty} U(z) = 0,$$

where $z = x - ct$ is the wave variable, $c \geq 0$ is the wave speed. Consider two cases: (a) without delay, i. e, (27); (b) with local delay, i. e., (26). We scale the model so that $d = 1$.

(a) Without Delay. Substitute $u(x,t) = U(z)$ into the reaction diffusion equation (27) without delay, i. e.,

$$\frac{\partial u}{\partial t} = \Delta u(t, x) - au(t, x) + b[1 - u(t, x)]u(t, x),$$

we obtain the traveling wave equation

$$U'' + cU' + (b - a - bU)U = 0,$$

which is equivalent to the following system of first order equations

$$U' = V,$$
$$V' = -cV - (b - a - bU)U. \tag{31}$$

System (31) has two equilibria: $E_0 = (0,0)$ and $E_1 = ((b - a)/b, 0)$. The following result shows that there is a traveling front solution of (31) connecting E_0 and E_1.

Theorem 12. *If $c \geq 2\sqrt{b-a}$, then in the (U,V) phase plane for system (31) there is a heteroclinic orbit connecting the critical points E_0 and E_1. The heteroclinic connection is confined to $V < 0$ and the traveling wave $U(z)$ is strictly monotonically decreasing.*

(b) With Local Delay. Consider the diffusive integro-differential equation (26) with a local delay kernel

$$G(t) = \frac{t}{\tau^2}\, \mathrm{e}^{-t/\tau}\,,$$

which is called the *strong* kernel. The parameter $\tau > 0$ measures the delay, which implies that a particular time in the past, namely τ time units ago, is more important than any other since the kernel achieves its unique maximum when $t = \tau$. Equation (26) becomes

$$\frac{\partial u}{\partial t} = \Delta u(t,x) - a\, u(t,x) + b[1 - u(t,x)] \int_{-\infty}^{t} \frac{t-s}{\tau^2}\, \mathrm{e}^{-\frac{t-s}{\tau}}\, u(s,x)\,\mathrm{d}s \quad (32)$$

for $(t,x) \in \mathbb{R}_+ \times \Omega$. Define $U(z) = u(x,t)$ and

$$W(z) = \int_0^\infty \frac{t}{\tau^2}\, \mathrm{e}^{-t/\tau} U(z + ct)\,\mathrm{d}t\,, \quad Y(z) = \int_0^\infty \frac{1}{\tau}\, \mathrm{e}^{-t/\tau} U(z + ct)\,\mathrm{d}t\,.$$

Differentiating with respect to z and denoting $U' = V$, we obtain the following traveling wave equations

$$\begin{aligned}
U' &= V\,,\\
V' &= aU - cV - bW + bUW\,,\\
c\tau W' &= W - Y\,,\\
c\tau Y' &= -U + Y\,.
\end{aligned} \qquad (33)$$

For $\tau > 0$, system (33) has two equilibria

$$(0,0,0,0) \quad \text{and} \quad \left(\frac{b-a}{b}, 0, \frac{b-a}{b}, \frac{b-a}{b}\right)\,.$$

A traveling front solution of the original equation exists if there exists a heteroclinic orbit connecting these two critical points.

Note that when τ is very small, system (33) is a singularly perturbed system. Let $z = \tau\eta$. Then system (33) becomes

$$\begin{aligned}
\dot{U} &= \tau V\,,\\
\dot{V} &= \tau(aU - cV - bW + bUW)\,,\\
c\dot{W} &= W - Y\,,\\
c\dot{Y} &= -U + Y\,,
\end{aligned} \qquad (34)$$

where dots denote differentiation with respect to η. While these two systems are equivalent for $\tau > 0$, the different time scales give rise to two different limiting systems. Letting $\tau \to 0$ in (33), we obtain

$$
\begin{aligned}
\dot{U} &= \tau V \,, \\
\dot{V} &= \tau(aU - cV - bW + bUW) \,, \\
0 &= W - Y \,, \\
0 &= -U + Y \,.
\end{aligned}
\tag{35}
$$

Thus, the flow of system (35) is confined to the set

$$
\mathcal{M}_0 = \{(U, V, W, Y) \in \mathbb{R}^4 : W = U, Y = U\}
\tag{36}
$$

and its dynamics are determined by the first two equations only. On the other hand, setting $\tau \to 0$ in (34) results in the system

$$
\begin{aligned}
U' &= 0 \,, \\
V' &= 0 \,, \\
cW' &= W - Y \,, \\
cY' &= -U + Y \,.
\end{aligned}
\tag{37}
$$

Any points in $\mathcal{M}_0$ are the equilibria of system (37). Generally, (33) is referred to as the *slow system* since the time scale z is slow, and (34) is referred to as the *fast system* since the time scale η is fast. Hence, U and V are called *slow variables* and W and Y are called the *fast variables*. $\mathcal{M}_0$ is the *slow manifold*.

If $\mathcal{M}_0$ is *normally hyperbolic*, then we can use the geometric singular perturbation theory of Fenichel (1979) to obtain a two-dimensional invariant manifold $\mathcal{M}_\tau$ for the flow when $0 < \tau \ll 1$, which implies the persistence of the slow manifold as well as the stable and unstable foliations. As a consequence, the dynamics in the vicinity of the slow manifold are completely determined by the one on the slow manifold. Therefore, we only need to study the flow of the slow system (33) restricted to $\mathcal{M}_\tau$ and show that the two-dimensional reduced system has a heteroclinic orbit.

Recall that $\mathcal{M}_0$ is a normally hyperbolic manifold if the linearization of the fast system (34), restricted to $\mathcal{M}_0$, has exactly $\dim \mathcal{M}_0$ eigenvalues with zero real part. The eigenvalues of the linearization of the fast system restricted to $\mathcal{M}_0$ are $0, 0, 1/c, 1/c$. Thus, $\mathcal{M}_0$ is normally hyperbolic.

The geometric singular perturbation theorem now implies that there exists a two-dimensional manifold $\mathcal{M}_\tau$ for $\tau > 0$. To determine $\mathcal{M}_\tau$ explicitly, we have

$$
\mathcal{M}_\tau = \{(U, V, W, Y) \in \mathbb{R}^4 : W = U + g(U, V; \tau), \quad Y = U + h(U, V; \tau)\} \,,
\tag{38}
$$

where the functions g and h are to be determined and satisfy

$$g(U, V; 0) = h(U, V; 0) = 0 \,.$$

By substituting into the slow system (33), we know that g and h satisfy

$$c\tau[(1 + \tfrac{\partial h}{\partial U} + \tfrac{\partial g}{\partial U})V + (\tfrac{\partial h}{\partial V} + \tfrac{\partial g}{\partial V})(aU - cV - b(U + h + g) \\ + bU(U + h + g))] = g \,,$$

$$c\tau[(1 + \tfrac{\partial h}{\partial U})V + \tfrac{\partial h}{\partial V}(aU - cV - b(U + h + g) \\ + bU(U + h + g))] = h \,.$$

Since h and g are zero when $\tau = 0$, we set

$$\begin{aligned} g(U, V; \tau) &= \tau g_1(U, V) + \tau^2 g_2(U, V) + \cdots \,, \\ h(U, V; \tau) &= \tau h_1(U, V) + \tau^2 h_2(U, V) + \cdots \,. \end{aligned} \tag{39}$$

Substituting $g(U, V; \tau)$ and $h(U, V; \tau)$ into the above equations and comparing powers of τ, we obtain

$$\begin{aligned} g_1(U, V) &= cV \,, \\ h_1(U, V) &= cV \,, \\ g_2(U, V) &= 2c^2(aU - cV - b(1 - U)U) \,, \\ h_2(U, V) &= c^2(aU - cV - b(1 - U)U) \,. \end{aligned} \tag{40}$$

The slow system (33) restricted to $\mathcal{M}_\tau$ is therefore given by

$$\begin{aligned} U' &= V \,, \\ V' &= aU - cV - b(1 - U)[U + g(U, V; \tau) + h(U, V; \tau)] \,, \end{aligned} \tag{41}$$

where g and h are given by (39) and (40). Note that when $\tau = 0$ system (41) reduces to the corresponding system (31) for the nondelay equation. We can see that for $0 < \tau \ll 1$ system (41) still has critical points E_0 and E_1. The following theorem shows that there is a heteroclinic orbit connecting E_0 and E_1 and thus equation (32) has a traveling wave solution connecting $u_0 = 0$ and $u_1 = (b - a)/b$.

Theorem 13 (Ruan and Xiao 2004). *For any $\tau > 0$ sufficiently small there exist a speed c such that the system (41) has a heteroclinic orbit connecting the two equilibrium points E_0 and E_1.*

The above results (Theorems 12 and 13) show that for the small delay the traveling waves are qualitatively similar to those of the non-delay equation. The existence of traveling front solutions show that there is a moving zone of transition from the disease-free state to the infective state.

Remark 4. When the delay kernel is non-local, for example,

$$F(x,t) = \frac{1}{\tau_0} e^{-\frac{t}{\tau_0}} \frac{1}{\sqrt{4\pi\rho_0}} e^{-\frac{x^2}{4\rho_0}}, \quad \tau_0 > 0, \quad \rho_0 > 0,$$

the existence of traveling wave solutions in (23) can be established by using the results in Wang, Li and Ruan (2006).

5.7 Discusion

Epidemic theory for homogeneous populations has shown that a critical quantity, known as the basic reproductive value (which may be considered as the fitness of a pathogen in a given population), must be greater than unity for the pathogen to invade a susceptible population (Anderson and May 1991). In reality, populations tend not to be homogeneous and there are nonlocal interactions. Therefore, there has been much theoretical investigation on the geographical spread of infectious diseases.

Invasion of diseases is now an international public health problem. The mechanisms of invasion of diseases to new territories may take many different forms and there are several ways to model such problems. One way is to introduce spatial effects into the model, divide the population into n subpopulations and allow infective individuals in one patch to infect susceptible individuals in another. The equilibrium behavior of such models has been studied widely, see Lajmanovich and Yorke (1976), Hethcote (1978), Dushoff and Levin (1995), Lloyd and May (1996), etc. It has been shown that spatial heterogeneity can reduce the occurrence of fade-outs in epidemic models (Bolker and Grenfell 1995).

Another way is to assume that there are nonlocal interactions between the susceptible and infective individuals and use integral equations to model the epidemics. In this short survey, we focused on the spatiotemporal dynamics of some nonlocal epidemiological models, include the classical Kermack–McKendrick model, the Kendall model given by differential and integral equations, the Diekmann–Thieme model described by a double integral equation, the diffusive integral equations proposed by De Mottoni et al. (1979) and Busenberg and Travis (1983), a vector-disease model described by a diffusive double integral equation (Ruan and Xiao 2004), etc.

For some diseases, such as vector-host diseases, the infectives at location x at the present time t were infected at another location y at an earlier time $t-s$. In order to study the effect of spatial heterogeneity (geographical movement), nonlocal interactions and time delay (latent period) on the spread of the disease, it is reasonable to consider more general models of the following form

$$\frac{\partial S}{\partial t} = d_1 \Delta S - S(x,t) \int_{-\infty}^{t} \int_{\Omega} I(y,s) K(x,y,t-s) \, dy \, ds,$$

$$\frac{\partial I}{\partial t} = d_2 \Delta I + S(x,t) \int_{-\infty}^{t} \int_{\Omega} I(y,s) K(x,y,t-s)\,\mathrm{d}y\,\mathrm{d}s - \gamma I(x,t)\,, \quad (42)$$

$$\frac{\partial R}{\partial t} = d_3 \Delta I + \gamma I(x,t)$$

under certain boundary and initial conditions, where d_1, d_2, d_3 are the diffusion rates for the susceptible, infective, and removed individuals, respectively. The kernel $K(x,y,t-s) \geq 0$ describes the interaction between the infective and susceptible individuals at location $x \in \Omega$ at the present time t which occurred at location $y \in \Omega$ at an earlier time $t-s$. It will be very interesting to study the spatiotemporal dynamics, such as stability of the disease-free equilibrium and existence of traveling waves, in the general model (42) and apply the results to study the geographical spread of some vector-borne diseases, such as West Nile virus and malaria.

Acknowledgement. I am grateful to Horst Thieme for giving me some very helpful comments and remarks on an earlier version of the paper and for bringing several references to my attention. I also would like to thank Odo Diekmann, Nick Britton and the referee for making some valuable comments and suggestions.

References

1. Anderson, R. M. and R. M. May (1991), *Infectious Diseases of Humans: Dynamics and Control,* Oxford University Press, Oxford.
2. Aronson, D. G. (1977), The asymptotic speed of propagation of a simple epidemic, in *"Nonlinear Diffusion",* eds. by W. E. Fitzgibbon and H. F. Walker, Research Notes in Math. **14**, Pitman, London, pp. 1–23.
3. Aronson, D. G. and H. F. Weinberger (1975), Nonlinear diffusion in population genetics, combustion, and nerve pulse propagation, in *"Partial Differential Equations and Related Topics",* ed. by J. A. Goldstein, Lecture Notes in Math. **446**, Springer, Berlin, pp. 5–49.
4. Aronson, D. G. and H. F. Weinberger (1975), Multidimennsional nonlinear diffusion arising in population genetics, *Adv. Math.* **30**: 33–76.
5. Atkinson, C. and G. E. H. Reuter (1976), Deterministic epidemic waves, *Math. Proc. Camb. Phil. Soc.* **80**: 315–330.
6. Barbour, A. D. (1977), The uniqueness of Atkinson and Reuter's epidemic waves, *Math. Proc. Camb. Phil. Soc.* **82**: 127–130.
7. Bartlett, M. S. (1956), Deterministic and stochastic models for recurrent epidemics, *Proc. 3rd Berkeley Symp. Math. Stat. Prob.* **4**: 81–109.
8. Bolker, B. M. and B. T. Grenfell (1995), Space, persistence, and dynamics of measles epidemics, *Phil. Trans. R. Soc. Lond.* **B237**: 298–219.
9. Brauer, F. and C. Castillo-Chavez (2000), *Mathematical Models in Population Biology and Epidemiology,* Springer-Verlag, New York.
10. Brown, K. J. and J. Carr (1977), Deterministic epidemic waves of critical velocity, *Math. Proc. Camb. Phil. Soc.* **81**: 431–433.
11. Busenberg, S. and K. L. Cooke (1978), Periodic solutions for a vector disease model, *SIAM J. Appl. Math.* **35**: 704–721.

12. Caraco, T., S. Glavanakov, G. Chen, J. E. Flaherty, T. K. Ohsumi and B. K. Szymanski (2002), Stage-structured infection transmission and a spatial epidemic: A model for lyme disease, *Am. Nat.* **160**: 348–359.

13. Capasso, V. (1978), Global solution for a diffusive nonlinear deterministic epidemic model, *SIAM J. Appl. Math.* **35**: 274–284.

14. Cliff, A., P. Haggett and M. Smallman-Raynor (1993), *Measles: An Historical Geography of a Major Human Viral Disease*, Blackwell, Oxford.

15. Cooke, K. L. (1979), Stability analysis for a vector disease model, *Rocky Mountain J. Math.* **9**: 31–42.

16. Cruickshank I., W. S. Gurney and A. R. Veitch (1999), The characteristics of epidemics and invasions with thresholds, *Theor. Pop. Biol.* **56**: 279–92.

17. De Monttoni, P., E. Orlandi and A. Tesei (1979), Aysmptotic behavior for a system describing epidemics with migration and spatial spread of infection, *Nonlinear Anal.* **3**: 663–675.

18. Diekmann, O. (1978), Thresholds and travelling waves for the geographical spread of infection, *J. Math. Biol.* **6**: 109–130.

19. Diekmann, O. (1979), Run for life: A note on the asymptotic spread of propagation of an epidemic, *J. Differential Equations* **33**: 58–73.

20. Diekmann, O. and J. A. P. Heesterbeek (2000), *Mathematical Epidemiology of Infective Diseases: Model Building, Analysis and Interpretation*, Wiley, New York.

21. Diekmann, O., J. A. P. Heesterbeek annd H. Metz (1995), The legacy of Kermack and McKendrick, in *"Epidemic Models: Their Structure and Relation to Data"*, ed. by D. Mollison, Cambridge Univ. Press, Cambridge, pp. 95–115.

22. Dushoff, J. and S. A. Levin (1995), The effects of population heterogeneityon disease invasion, *Math. Biosci.* **128**: 25–40.

23. Fenichel, N. (1979), Geometric singular perturbation theory for ordinary differential equations. *J. Differential Equations* **31**: 53–98.

24. Fisher, R. A. (1937), The wave of advance of advantageous genes, *Ann. Eugenics* **7**: 353–369.

25. Grenfell, B. T., O. N. Bjornstad and J. Kappey (2001), Travelling waves and spatial hierarchies in measles epidemics, *Nature* **414**: 716–723.

26. Hethcote, H. W. (2000), The mathematics of infectious disease, *SIAM Rev.* **42**: 599–653.

27. Hosono, Y. and B. Ilyas (1995), Traveling waves for a simple diffusive epidemic model, *Math. Models Methods Appl. Sci.* **5**: 935–966.

28. Källen, A., P. Arcuri and J. D. Murray (1985), A simple model for the spatial spread and control of rabies, *J. Theor. Biol.* **116**: 377–393.

29. Kendall, D. G. (1957), Discussion of 'Measles periodicity and community size' by M. S. Bartlett, *J. Roy. Stat. Soc.* **A120**: 64–76.

30. Kendall, D. G. (1965), Mathematical models of the spread of injection, in *"Mathematics and ComputerScience in Biology and Medicine"*, Medical Research Council, London, pp. 213–225

31. Kermack, W. O. and A. G. McKendrik (1927), A contribution to the mathematical theory of epidemics, *Proc. Roy. Soc.* **A115**: 700–721.

32. Kolmogorov, A., I. Petrovskii, and N. S. Piskunov (1937), Étude de l'équation de le diffusion avec croissance de la quantité de mateère et son application à un problème biologique, *Moscow Univ. Bull.* **1**: 1–25.

33. Lajmanovich, A. and J. A. Yorke (1976), A deterministic model for gonorrhea in a non-homogeneous population, *Math. Biosci.* **28**: 221–236.

34. Lloyd, A. L. and R. M. May (1996), Spatial heterogeneity in epidemic models, *J. Theor. Biol.* **179**: 1–11.
35. Lounibos, L. P. (2002), Invasions by insect vectors of human disease, *Ann. Rev. Entomol.* **47**: 233–266.
36. Marcati, P. and M. A. Pozio (1980), Global asymptotic stability for a vector disease model with spatial spread, *J. Math. Biol.* **9**: 179–187.
37. Medlock, J. and M. Kot (2003), Spreding disease: integro-differential equations old and new, *Math. Biosci.* **184**: 201–222.
38. Mollison, D. (1972), Possible velocities for a simple epidemic, *Adv. Appl. Prob.* **4**: 233–257.
39. Mollison, D.(1991), Dependence of epidemic and population velocities on basic parameters, *Math. Biosci.* **107**: 255–87.
40. Murray, J. D. (1989), *Mathematical Biology*, Springer-Verlag, Berlin.
41. Murray, J. D. and W. L. Seward (1992), On the spatial spread of.rabies among foxes with immunity, *J. Theor. Biol.* **156**: 327–348.
42. Murray, J. D., E. A. Stanley and D. L. Brown (1986), On the spatial spread of rabies among foxes, *Proc. R. Soc. Lond.* **B229**(1255): 111–150.
43. O'Callagham, M. and A. G. Murray (2002), A tractable deterministic model with realistic latent periodic for an epidemic in a linear habitat, *J. Math. Biol.* **44**: 227–251.
44. Pozio, M. A. (1980), Behavior of solutions of some abstract functional differential equations and application to predator-prey dynamics, *Nonlinear Anal.* **4**: 917–938.
45. Pozio, M. A. (1983), Some conditions for global asymptotic stability of equilibria of integrodifferential equations, *J. Math. Anal. Appl.* **95**: 501–527.
46. Rass, L. and J. Radcliffe (2003), *Spatial Deterministic Epidemics*, Math. Surveys Monogr. **102**, Amer. Math. Soc., Providence.
47. Ross, R. and H. P. Hudson (1917), Ann appliicationn of the theory of probability to the study of a priori pathometry - Part III, *Proc. R. Soc. Lond.* **A93**: 225–240.
48. Ruan, S. and J. Wu (1994), Reaction-diffusion equations with infinite delay, *Can. Appl. Math. Quart.* **2**: 485–550.
49. Ruan, S. and D. Xiao (2004), Stability of steady states and existence of traveling waves in a vector disease model, *Proc. Royal Soc. Edinburgh Sect.* **A134**: 991–1011.
50. Schumacher, K. (1980a), Travelling-front solution for integro-differential equation. I, *J. Reine Angew. Math.* **316**: 54–70.
51. Schumacher, K. (1980b), Travelling-front solution for integro-differential equation. II, in *"Biological Growth and Spread"*, Lecture Notes in Biomathematics **38**, eds. by W. Jäger, H. Rost and P. Tautu, Springer, Berlin, pp. 296–309.
52. Thieme, H. R. (1977a), A model for the spatial spread of an epidemic, *J. Math. Biol.* **4**: 337–351.
53. Thieme H. R. (1977b), The asymptotic behavior of solutions of nonlinear integral equations, *Math. Zeitchrift* **157**: 141–154.
54. Thieme, H. R. (1979), Asymptotic estimates of the solutions of nonlinear integral equation and asymptotic speeds for the spread of populations, *J. Reine Angew. Math.* **306**: 94–121.
55. Thieme, H. R. (1980), Some mathematical considerations of how to stop the spatial spread of a rabies epidemics, in *"Biological Growth and Spread,"* Lecture

Notes in Biomathematics **38**, eds. by W. Jäger, H. Rost and P. Tautu, Springer, Berlin, pp. 310–319.

56. Thieme, H. R. (1985), Renewal theorems for some mathematical models in epidemiology, *J. Integral Equations* **8**: 185–216.

57. Thieme, H. R. (1988), Asymptotic proportionality (weak ergodicity) and conditional asymptotic equality of solutions to time-heterogeneous sublinear difference and differential equations, *J. Differential Equations* **73**: 237–268.

58. Thieme, H. R. (2003), *Mathematics in Population Biology*, Princeton University Press, Princeton.

59. Thieme, H. R. and X.-Q. Zhao (2003), Asymptotic speeds of spread and traveling waves for integral equations and delayed reaction-diffusion models, *J. Differential Equations* **195**: 430–470.

60. Van den Bosch, F., J. A. J. Metz and O. Diekmann (1990), The velocity of spatial population expansion, *J. Math. Biol.* **28**: 529–565.

61. Volz, R. (1982), Global asymptotic stability of a periodic solution to an epidemic model, *J. Math. Biol.* **15**: 319–338.

62. Wang, Z.-C., W.-T. Li and S. Ruan (2006), Traveling wave fronts in reaction-diffusion systems with spatio-temporal delays, *J. Differential Equations*, **222**: 185–232.

63. Webb, G. F. (1981), A reaction-diffusion model for a deterministic diffusive epidemic, *J. Math. Anal. Appl.* **84** (1981), 150–161.

64. Wu, J. (1996), *Theory and Applications of Partial Functional Differential Equations*, Springer, New York.

6

Pathogen Competition and Coexistence and the Evolution of Virulence [*]

Horst R. Thieme

Summary. Competition between different strains of a micro-parasite which provide complete cross-protection and cross-immunity against each other selects for maximal basic replacement ratio if, in the absence of the disease, the host population is exclusively limited in its growth by a nonlinear population birth rate. For mass action incidence, the principle of $\mathcal{R}_0$ maximization can be extended to exponentially growing populations, if the exponential growth rate is small enough that the disease can limit population growth. For standard incidence, though not in full extent, it can be extended to populations which, without the disease, either grow exponentially or are growth-limited by a nonlinear population death rate, provided that disease prevalence is low and there is no immunity to the disease. If disease prevalence is high, strain competition rather selects for low disease fatality. A strain which would go extinct on its own can coexist with a more virulent strain by protecting from it, if it has strong vertical transmission.

6.1 Introduction

With their rapid turn-over, parasitic populations are ideal objects to study the pattern and processes of evolution (evolution of virulence, co-evolution of hosts and parasites) (Anderson and May 1982; ;Bull et al. 1991; Clayton and Moore 1997; Lenski and Levin 1985; Levin 1996; Levin and Lenski 1985; Levin et al. 1977). In turn, it is important to understand these principles in order to control infectious diseases without creating resistant parasites or drive them towards increased virulence (virulence management) (Dieckmann et al. 2002; Ewald 1994; Levin 1999; Stearns 1999). Mathematical modeling and model analysis are very much needed for deeper understanding and as a theoretical laboratory to devise control and management strategies. Evolution in host-pathogen systems in general and evolution of virulence in particular have been extensively researched in the last 20 or 25 years. An excellent compilation has appeared not too long ago (Dieckmann et al. 2002), with

[*] partially supported by NSF grant 0314529

50 pages of references and an index of 17 pages. Many theoretical studies on the evolution of virulence start from the assumption that, in the absence of multiple infections and the presence of complete cross-immunity, only the parasite strain persists that has the maximal basic replacement ratio (Anderson and May 1982, 1992; Day 2002a, 2002b; Dieckmann et al. 2002) and the references therein). (In this paper I prefer *replacement ratio* over *reproduction number* because we reserve the second for the host population which has its own dynamics with births and deaths). This assumption (also called $\mathcal{R}_0$ maximization) has been found to be valid under quite a few restrictions: the density-dependent regulation of the host occurs through the birth rate, and the infection is horizontally transmitted and does not lower the fertility (Bremermann and Thieme 1989; Castillo-Chavez and Thieme 1995; Castillo-Chavez et al. 1996). The general validity of such a principle has recently been challenged on theoretical grounds (Dieckmann 2002a; Metz et al. 1996), because the basic replacement ratio only describes the performance of a strain in an almost infection-free host population but not its capacity of invading a host population already infected with other strains. Indeed, in spite of complete cross-protection and complete cross-immunity, coexistence of several strains in one host population has been found in the following scenarios:

- The infection occurs both vertically and horizontally (Lipsitch et al. 1996). Simulations show that strains with lower virulence can outcompete strains with higher basic replacement ratio.
- The density-dependent regulation of the host occurs through the death rate rather than the birth rate (Andreasen and Pugliese 1995; Ackleh and Allen 2003, 2005).
- There is no density-dependent regulation of the host at all such that the host would grow exponentially in the absence of the disease, and the incidence is of standard type (Lipsitch and Nowak 1995a).

Coexistence of different strains does not only violate the principle of $\mathcal{R}_0$ maximization (typically the coexisting strains will have different basic replacement ratios), but also the principle of competitive exclusion which would allow only one consumer (the parasite) to live on a single resource (the host) (Hutchinson 1978; Levin 1970; May 2001). Coexistence of different strains is less surprising if the host population is structured in one way or another effectively creating several resources for the parasite as in the following scenarios:

- The disease is heterosexually transmitted with differential female susceptibility (Castillo-Chavez et al. 1999).
- The disease is homosexually transmitted between and within two host groups with different responses to the disease (Li et al. 2003).

The picture becomes more complicated under super-infection and co-infection (Levin and Pimentel 1981; Castillo-Chavez 2002b; Feng and Velasco-Hernández 1997; Castillo-Chavez and Velasco-Hernández 1998; Dieckmann et al. 2002b; Iannelli 2005; Pugliese 2002b; Esteva and Vargas 2003; Tanaka and

Feldman 1999; Day and Proulx 2004 and references therein), under incomplete cross-immunity (Andreason 1997; Andreason et al. 1997; Gog and Swinton 2002; Lin et al. 1999; Nuño et al. to appear; Dawns and Gog 2002), or under (frequent) mutation, recombination, or within-host evolution of the parasite (Dieckmann et al. 2002b; Martcheva et al.; Regoes et al. 2000; Tanaka and Feldman 1999; Day and Proulx 2004 and references therein). The competition between several strains is also influenced by host population structure in the form of partnerships, kinships, or other social structures (Dieckmann et al. 2002b and references therein), by spatial structure in the form of nearest neighborhood infection or patchiness (Haraguchi and Sasaki 2000; Charles et al. 2002), community structure (Bowers and Turner 1997), or, in macroparasitic diseases, by the mode of parasite reproduction (Pugliese 2002a).

In this chapter, we will extend the conditions under which strain competition will select the maximal basic replacement ratio, we will determine under which conditions something else (actually what?) may be maximized (in both cases with competitive exclusion of sub-optimal strains), and under which conditions different strains may coexist. For mathematical manageability, we make the assumption that infection with one strain infers complete cross-protection against infection with other strains during the infection period and complete immunity and cross-immunity during the recovery period (if there is any).

6.2 Host populations with nonlinear birth rates and arbitrary incidence

We consider a host population the size of which, N, develops according to the differential equation $N' = B(N) - \mu N$ in the absence of the infectious disease. Here $B(N)$ is the population birth rate and μ the (constant) per capita mortality rate. This means that the population growth is only regulated by the nonlinear birth rate.

Enters the parasite which, through mutation, can occur in several, n, strains. Mutation is assumed so rare, that it is not included in the mathematical model which describes the competition of the strains in the host population without further mutations. The selection process is assumed to be essentially terminated (i. e. the model dynamics have settled at their large time behavior) when new mutations occur creating a new ensemble of competing strains to which the mathematical model again applies.

Variables and parameters

Independent variables

n	total number of strains
j, k	strain indices
$t \geq 0$	time

Dependent variables

N	total population size
I_j	number of infectives with strain j
v_j	proportion of infectives with strain j

Demographic parameters

$\beta > 0$	per capita birth rate of susceptibles
$B(N)$	population birth rate at population size N
$\mu > 0$	per capita mortality rate of susceptibles

Epidemic parameters

$p_j \in [0, 1)$	fraction of horizontal transmission of strain j
$q_j \in [0, 1]$	ratio of fertility in strain j infectives to susceptibles
$\eta_j > 0$	per capita/capita infection probability (or rate) for strain j
$\gamma_j \geq 0$	per capita recovery rate of strain j infectives
$\alpha_j \geq 0$	disease death rates of strain j infectives

For $j = 1, \ldots, n$, I_j denotes the number of infective hosts which are infected by strain j. The host equation must be modified to take account of disease fatalities, α_j is the extra per capita mortality rate of infectives due to strain j. Infection with one strain is assumed to protect completely against the infection with other strains,

$$\left.\begin{aligned}
N' &\leq B(N) - \mu N - \sum_{j=1}^{n} \alpha_j I_j \,, \\
I'_j &= \tilde{C}(t)\eta_j I_j - (\mu + \alpha_j + \gamma_j)I_j \,, \quad j = 1, \ldots, n \,, \\
I_j(t) &\geq 0 \,, \quad j = 1, \ldots, n \,; \quad \sum_{j=1}^{n} I_j(t) \leq N \,.
\end{aligned}\right\} \tag{1}$$

The host equation is formulated as an inequality because there may be a class of recovered individuals which is not explicitly modeled. The inequality allows for the recovered individuals to have an increased mortality as well, and for infective and/or recovered individuals to have a reduced birth rate. From a mathematical point of view, the inequality prohibits the model from being closed such that a discussion of existence and uniqueness of solutions is not meaningful, but the model is complete enough for our purposes. The function $\tilde{C}$ gives the per capita rate of contacts with susceptible individuals. It certainly depends on the number of susceptible individuals, but may also depend on the total population size or the sizes of other epidemiologic classes (Castillo-Chavez and Thieme 1995) and may arbitrarily depend on time. In this context, η_j is the average probability at which the contact between a susceptible and infective individual leads to an infection. γ_j is the per capita rate of recovering from an infection with strain j. The recovery may lead to permanent or temporary immunity or directly back into the susceptible class; there is complete cross-immunity in the sense that individuals that are permanently or temporarily immune against one strain are also immune against all other strains. The fraction $\frac{1}{\mu+\alpha_j+\gamma_j}$ is the average sojourn time in the infectious stage if one is infected by strain j (including that the sojourn may be cut short by natural or disease death). The fraction $\frac{\alpha_j}{\mu+\alpha_j+\gamma_j}$ is the probability to die from a strain j infection during the infectious period (Thieme 2003). We define the *relative replacement ratio* of strain j as

$$\mathcal{R}_j = \frac{\eta_j}{\mu + \alpha_j + \gamma_j} . \tag{2}$$

If $\tilde{C}$ did not depend on time and gave the per capita contact rate with all individuals in the population, $\tilde{C}\mathcal{R}_j$ were the *basic replacement ratio* (basic reproduction number) of strain j, the average number of secondary infections caused by one typical strain j infective in an otherwise completely susceptible population (Brauer and Castillo-Chavez 2001; Diekmann and Heesterbeek 2000; Dietz 1975; Anderson and May 1991; Hethcote 2000; Thieme 2003).

Notice that the fraction $\frac{\mathcal{R}_j}{\mathcal{R}_k}$ equals the fraction of the basic replacement ratio of strain j over the basic replacement ratio of strain k in case that meaningful basic replacement ratios can be defined. The next result states that a strain dies out if there is another strain with a higher relative replacement ratio.

Theorem 1. *Let* $\limsup_{N\to\infty} \frac{B(N)}{N} < \mu$. *If* $\mathcal{R}_n < \mathcal{R}_j$ *for some* $j < n$, *then* $I_n(t) \to 0$ *as* $t \to \infty$.

Since we can renumber the strains as needed, we obtain that the strain with maximal relative replacement ratio drives all other strains into extinction.

Corollary 1. *Let* $\limsup_{N\to\infty} \frac{B(N)}{N} < \mu$. *If* $\mathcal{R}_j < \mathcal{R}_1$ *for all* $j = 2,\ldots,n$, *then* $I_j(t) \to 0$ *for* $t \to \infty$, $j = 2,\ldots,n$.

The theorem can be proved as in Saunders (1981), Bremermann and Thieme (1989), or Castillo-Chavez and Thieme(1995). Here we use the method proposed by Ackleh and Allen (2003; 2005) which is more flexible.

A Lyapunov type selection functional

We fix j and set

$$y = I_n^{\xi_n} I_j^{-\xi_j} ,$$

with $\xi_j > 0$ to be determined later. Then

$$\frac{y'}{y} = \xi_n \frac{I_n'}{I_n} - \xi_j \frac{I_j'}{I_j} .$$

128   Horst R. Thieme

We substitute the differential equations for I_j in (1),

$$\frac{y'}{y} = \xi_n\Big(\tilde{C}(t)\eta_n - (\mu_n + \alpha_n + \gamma_n)\Big) - \xi_j\Big(\tilde{C}(t)\eta_j - (\mu_j + \alpha_j + \gamma_j)\Big) \, .$$

Proof. To prove Theorem 1, we choose $\xi_j = \frac{1}{\eta_j}$. By (2),

$$\frac{y'}{y} = \frac{1}{\mathcal{R}_j} - \frac{1}{\mathcal{R}_n} < 0 \, .$$

This implies that $y(t) \to 0$ as $t \to \infty$. From the definition of y,

$$I_n(t)^{\xi_n} \le y(t) I_j(t)^{\xi_j} \le y(t) N(t)^{\xi_j} \, .$$

The assumption for $B(N)$ and the differential inequality for N in (1) imply that N is bounded and so $I_n(t) \to 0$ as $t \to \infty$.

Corollary 1 only states that all strains but the one with maximal relative replacement ratio go extinct; if the model is closed in an appropriate way, one can show that the strain with maximal relative replacement ratio persists provided that its basic replacement ratio exceeds 1 (Bremermann and Thieme 1989).

6.2.1 Propagule producing parasites

The model (1) is quite general as far as the disease incidence is concerned, but restrictive as we assume that the growth of the host population is bounded by a nonlinear population birth rate while the population death rate is linear. It is further restrictive, as the mathematical technique does not work if several infectious stages or a latent period are incorporated into the model or other features are added. As an illustration, we revisit a model (Ewald and De Leo 2002) for an infectious disease, which is spread both by direct (horizontal) transmission and by waterborne (or otherwise free-living) propagules released by infective hosts. As additional dependent variables, we add the amount of waterborne propagules, W_j, released by infectives with strain j. Recast in the notation used above, the model in Ewald and De Leo (2002) takes the form

$$\left.\begin{aligned}
N' &= B(N) - \mu N - \sum_{j=1}^{n} \alpha_j I_j \, , \\
I_j' &= \tilde{C}(t)\big(\hat{\eta}_j I_j + \tilde{\eta}_j W_j\big) - (\mu + \alpha_j + \gamma_j)I_j \, , \\
W_j' &= \sigma_j I_j - \nu_j W_j \, ,
\end{aligned}\right\} \qquad j = 1, \ldots, n, \qquad (3)$$

where $B(N) = B$ is constant and $\tilde{C} = S = N - \sum_{j=1}^{n} I_j$ is the number of susceptible individuals. The parameters σ_j are the per capita release rates of waterborne propagules and ν_j are their per unit destruction rates, for strain j.

$\hat{\eta}_j$ is the direct horizontal per capita/capita infection rate, while $\tilde{\eta}$ is the infection rate per capita host and per unit propagule. So far, all attempts have failed to modify the proof of Theorem 1 such that the addition of waterborne pathogens is included. In view of the counterexamples of $\mathcal{R}$ maximization cited in the Introduction and presented below, the mathematical underpinning of the interesting results in Ewald and De Leo (2002) still needs to be provided. However, if the parameters σ_j and ν_j are large compared with the other parameters, in other words, if the dynamics of waterborne pathogens are fast compared with the dynamics of the remaining system, a quasi-steady state approximation may be justified for the waterborne pathogens,

$$W_j \approx \frac{\sigma_j}{\nu_j} I_j \, .$$

If we substitute this relation as an equality into the equations for I_j in (3), we obtain (1) with

$$\eta_j = \hat{\eta}_j + \tilde{\eta}_j \frac{\sigma_j}{\nu_j} \, ,$$

and Corollary 1 applies. This means that the parasite evolves in such a way that $\mathcal{R}$ is increased.

6.2.2 An example of virulence management

Much of the theoretical consideration as to whether parasite evolution leads to low, intermediate, or high virulence is based on the idea that the mutations a parasite may undergo are subject to certain constraints. If a mutation leads to a higher replication rate of the parasite, the greater exploitation of host resources will presumably increase the harm done to the host resulting in decreased mobility and/or fecundity and increased disease mortality. The relationship of these trade offs, which crucially depends on the specifics of the disease, has been considered in many studies (Dieckmann et al. 2002; Anderson and May 1982; Anderson and May 1991; Day 2001; Day 2002b; Ganusov et al. 2002; Davies et al. 2000; Brunner 2004 and references therein). If one takes a parasite-centered rather than a host-centered point of view, one can make the case that the impact of waterborne pathogens is possibly even more dramatic than presented in (Ewald and De Leo 2002; Day 2002a). Let ζ be the replication rate of parasites in an infectious host. Then $\hat{\eta} = \zeta f(\zeta)$ where the decreasing function f takes account of the degree by which the infective individual looses its mobility. If the disease has a diarrhetic component, insufficient hygiene leads to waterborne transmission. We assume that $\tilde{\eta}\frac{\sigma}{\nu} = \kappa_1 \zeta$. We assume that the disease mortality rate responds in a monotone increasing way to an increase in the replication rate, though we assume that this response is weak on the one hand, but stronger than linear on the other hand

$$\alpha = \kappa_2 \zeta^b \, ,$$

with $b > 1$, and κ_2 being small compared with the other parameters. It is difficult to assess how the replication rate influences the recovery rate γ. Ewald and De Leo (2002) assume a decreasing relationship (as has been found for myxomatosis (Anderson and May 1991)), but one can also imagine that a high replication rate triggers a strong immune response such that, while the length of morbidity may increase, the duration of the actual infectious period may decrease. We assume for simplicity that γ does not depend on ζ. So

$$\mathcal{R} = \frac{\zeta f(\zeta) + \kappa_1 \zeta}{\mu + \gamma + \kappa_2 \zeta^b} \, .$$

We choose $f(\zeta) = \kappa_3 e^{-\kappa_4 \zeta}$. After some scaling, one can assume that $\kappa_3 = \kappa_4 = 1$ and $\mu + \gamma = 1$, and, without loosing any generality, we can consider

$$\mathcal{R}(\zeta, \kappa) = \frac{\zeta e^{-\zeta} + \kappa \zeta}{1 + \epsilon \zeta^b} \, ,$$

with suitable compound parameters $\epsilon, \kappa > 0$. The weak response of the disease death rate to an increase in the replication rate means that $\epsilon > 0$ is small. It is more instructive to write $\mathcal{R}$ as a linear combination $\mathcal{R}(\zeta, \kappa) = h(\zeta) + \kappa g(\zeta)$ with

$$h(\zeta) = \frac{\zeta e^{-\zeta}}{1 + \epsilon \zeta^b} \, , \quad g(\zeta) = \frac{\zeta}{1 + \epsilon \zeta^b} \, .$$

The compound parameter $\kappa \geq 0$ is proportional to the replication rate of the parasite, to the release rate of waterborne propagules into the water and to their survival time. In particular, it reflects the quality of hygiene in inverse proportion: $\kappa = 0$ means perfect hygiene. Both h and g are uni-modal functions, i. e. they first increase and then decrease. h takes it maximum at some value $\zeta_h(\epsilon) < 1$ and the maximum is less than e^{-1}. g takes is maximum at $\zeta_g(\epsilon) = (\epsilon(b-1))^{-1/b}$ and the maximum is $b^{-1} \epsilon^{-1/b} (b-1)^{1-(1/b)}$. So both the maximum of g and the argument where g takes its maximum tend to infinity as $\epsilon \to 0$. Under perfect hygiene, $\kappa = 0$, evolution will lead to a replication rate less than the (scaled) value 1 and a per capita disease death rate less than ϵe^{-b}, while for small ϵ and imperfect hygiene, $\kappa > 0$, it will lead to a replication rate much higher than 1 and also to a substantially larger disease death rate. One can even construct a relative replacement ratio which has two local maxima for appropriate κ.

6.3 Host populations with linear birth rates

We turn to host populations which, like the human population, are assumed to grow exponentially in the time frame of epidemiologic relevance. For mathematical managability, we restrict our consideration to diseases which only involve susceptible and infective individuals. Since the per capita birth rates are constant, it is not too complicated to incorporate vertical disease transmission and fertility reduction of infectious individuals. The per capita birth rate of a susceptible individual is denoted by β, the per capita birth rate of a strain j infective by $q_j\beta$ with $q_j \in [0,1]$, and the proportion of the offspring of a strain j infective which is also infective (with the same strain) by p_j. $1 - q_j$ is the fraction by which the fertility of a strain j infective is reduced compared with a susceptible individual and is one of the measures of the strain's virulence. At this point, we formulate a model with a general incidence which we will specify later. Let $C(N)$ be the rate at which a single individual makes contacts in a population of size N. In this context, let η_j be the probability that a contact between a susceptible individual and an infective individual of strain j leads to an infection. Since I_j/N is the probability that an average contact actually occurs with an infective individual with strain j, the incidence of strain j infectives is given by $\frac{C(N)}{N}S\eta_j I_j$, where $S = N - \sum_{k=1}^{n} I_k$ is the number of susceptible individuals. The population birth rate is $\beta S + \sum_{k=1}^{n} q_k \beta I_k = \beta N - \sum_{k=1}^{n} \beta(1-q_k)I_k$. Though we will first consider a natural per capita mortality rate μ which does not depend on the population size N, we allow for a size-dependent per capita mortality rate $\mu(N)$ for later purposes. Including disease fatalities, the population mortality rate is $\mu(N)N + \sum_{k=1}^{n} \alpha_k I_k$. The model equations are these,

$$\left.\begin{aligned}
N' &= (\beta - \mu(N))N - \sum_{k=1}^{n}(\alpha_k + \beta(1-q_k))I_k\,, \\
I'_j &= \beta q_j p_j I_j + \frac{C(N)}{N}\left(N - \sum_{k=1}^{n} I_k\right)\eta_j I_j - (\mu(N) + \alpha_j + \gamma_j)I_j\,, \\
&\quad j = 1,\ldots,n\,.
\end{aligned}\right\} \tag{4}$$

It is also informative to derive the equations for the proportions of strain j infectives, v_j, and the proportion of all infectives, v,

$$v_j = \frac{I_j}{N}\,, \quad v = \sum_{k=1}^{n} v_j\,.$$

By the quotient rule,

$$v'_j = \frac{I'_j}{N} - v_j\frac{N'}{N}\,.$$

By (4),

$$\left.\begin{aligned}
N' &= N\left(\beta - \mu(N) - \sum_{k=1}^{n} b_k v_k\right)\,, \\
v'_j &= v_j\left(C(N)(1-v)\eta_j - a_k + \sum_{k=1}^{n} b_k v_k\right)\,.
\end{aligned}\right\} \tag{5}$$

Here
$$a_j = \alpha_j + \gamma_j + \beta(1 - q_j p_j)\,, \quad b_j = \alpha_j + \beta(1 - q_j)\,. \tag{6}$$
Notice that $a_j \geq b_j \geq 0$. We add the equations for v_j in (5),
$$v' = C(N)(1 - v)\sum_{k=1}^{\infty} \eta_k v_k - \sum_{k=1}^{n} a_k v_k + v\sum_{k=1}^{n} b_k v_k\,. \tag{7}$$
Since $a_j \geq b_j$,
$$v' \leq (1 - v)\sum_{k=1}^{\infty}\bigl(C(N)\eta_k - a_k\bigr)v_k\,, \tag{8}$$
with the inequality being strict for $v > 0$. In accordance with the interpretation of v_j we have the following result from this inequality.

Proposition 1. *If $v_j(t) \geq 0$ and $v(t) \leq 1$ holds for $t = 0$, then it holds for all $t > 0$.*

6.4 Exponential growth and mass action incidence

Different strains can coexist, if the host population growth exponentially and the contact rate in a population of size N, $C(N)$, is independent of N (*standard incidence*) (Lipsitch, Nowak 1995). We will derive a condition for competitive exclusion between different strains under the *mass action incidence law* which assumes that $C(N)$ is proportional to N. We absorb the proportionality factor into the parameter η_j such that $C(N) = N$. η_j must now be interpreted as a per capita/capita infection rate, i.e., $\frac{1}{\eta_j}$ is the average time it takes for a typical infective and a typical susceptible individual to contact each other and to transmit the disease. We assume that the population mortality rate is linear (the per capita mortality rate does not depend on population size) and system (4) specializes to
$$\left.\begin{aligned}
N' &= \beta N - \mu N - \sum_{k=1}^{n}\bigl(\alpha_k + \beta(1 - q_k)\bigr)I_k\,, \\
I'_j &= \eta_j I_j\Bigl(N - \sum_{k=1}^{n} I_k - \mathcal{R}_j^{-1}\Bigr)\,, \quad j = 1,\dots,n\,,
\end{aligned}\right\} \tag{9}$$
with
$$\mathcal{R}_j^{-1} := \frac{\mu + \alpha_j + \gamma_j - \beta q_j p_j}{\eta_j}\,. \tag{10}$$
This system is the multi-strain version of a model in Busenberg et al. (1983) (see also Busenberg and Cooke 1993). The notation $\mathcal{R}_j^{-1}$ is motivated by the fact that it is the reciprocal of the relative replacement ratio in (2) if there

is no vertical transmission, $p_j = 0$. If the proportion of vertical transmission is so small that $\mathcal{R}_j^{-1} > 0$ for $j = 1, \ldots, n$,

$$\mathcal{R}_j = \frac{\eta_j}{\mu + \alpha_j + \gamma_j - \beta q_j p_j} \tag{11}$$

can be understood as a relative replacement ratio that is amplified by vertical transmission. Recall that

$$D_j = \frac{1}{\mu + \alpha_j + \gamma_j}$$

is the expected sojourn in the strain j infective stage. Then

$$\tilde{D}_j = \frac{1}{\mu + \alpha_j + \gamma_j - \beta q_j p_j} = D_j \frac{1}{1 - D_j \beta q_j p_j} = D_j \sum_{m=0}^{\infty} (D_j \beta q_j p_j)^m$$

is the *expected cumulative sojourn in the infective stage* of a strain j infective individual and all its vertically produced generations. Therefore we call $\mathcal{R}_j$ in (11) the *(vertically) cumulative relative replacement ratio* of strain j provided that $\mathcal{R}_j^{-1} > 0$. However, in order to include the case in which some or all $\mathcal{R}_j^{-1}$ are negative or zero, we will formulate the forthcoming results in terms of $\mathcal{R}_j^{-1}$ rather than $\mathcal{R}_j$.

In terms of the proportions of infectives with strain j, (5) specializes to

$$\left. \begin{aligned} N' &= N\left(\beta - \mu - \sum_{k=1}^{n} b_k v_k\right), \\ v_j' &= v_j\left(N(1 - v)\eta_j - a_j + \sum_{k=1}^{n} b_k v_k\right), \end{aligned} \right\} \tag{12}$$

and (7) to

$$v' = N(1 - v)\sum_{k=1}^{\infty} \eta_k v_k - \sum_{k=1}^{n} a_k v_k + v \sum_{k=1}^{n} b_k v_k . \tag{13}$$

Recall the definitions of a_j and b_j in (6). *Throughout this section, we assume without further mentioning* that, for each j, $a_j - b_j = \gamma_j + \beta q_j(1 - p_j) > 0$. This means that each strain j has a positive recovery rate or both imperfect vertical transmission and imperfect sterilization by infection. We further assume that $\beta > \mu$ such that the population increases exponentially in the absence of the disease.

Proposition 2. *The parasite population is uniformly weakly persistent in the sense that there exists some $\epsilon > 0$ such that $\limsup_{t\to\infty} \sum_{j=1}^{n} \frac{I_j(t)}{N(t)} \geq \epsilon$ for all solutions of (9) with $N(0) > 0$, $\sum_{j=1}^{n} I_j(0) > 0$.*

The host population is strongly uniformly persistent in the sense that there exists some $\epsilon > 0$ such that $\liminf_{t\to\infty} N(t) \geq \epsilon$ for all solutions of (9) with $N(0) > 0$.

Proof. For $v = \sum_{j=1}^{n} \frac{I_j}{N}$,

$$\frac{N'}{N} \geq \beta - \mu - (\max_{k=1}^{n} b_k)v ,$$

$$\frac{v'}{v} \geq N(1-v) \min_{k=1}^{n} \eta_k - \max_{k=1}^{n} a_k .$$

Choose $\epsilon \in (0,1)$ such that $\beta - \mu - \epsilon \max_{k=1}^{n} b_k > 0$. Assume $\limsup_{t\to\infty} v(t) < \epsilon$. Then $N(t) \to \infty$ as $t \to \infty$ by the first equation and $v(t) \to 1$ as $t \to \infty$ by the second equation, a contradiction. Hence there exists an $\epsilon > 0$ such that $\limsup_{t\to\infty} v(t) \geq \epsilon$ for all solutions with $N(0) > 0$ and $v(0) > 0$.

As for the host population, we have that $N(t) = N(0) \, e^{(\beta - \mu)t} \to \infty$ as $t \to \infty$, if $v = \sum_{j=1}^{n} \frac{I_j}{N}$ is zero at $t = 0$, because then $v(t) = 0$ for all $t \geq 0$. Let us assume that $v(0) > 0$ and that $\limsup_{t\to\infty} N(t) < \epsilon$ for some $\epsilon > 0$. Set $\gamma_\diamond = \min_k(\gamma_k + \beta q_k(1 - p_k)) > 0$, $\eta^\diamond = \max_k \eta_k$. Since $v \leq 1$ by Proposition 1, it follows from (13) that

$$\frac{v'}{v} \leq N\eta^\diamond - \gamma_\diamond .$$

By our assumption, if we choose $\epsilon > 0$ small enough,

$$\limsup_{t\to\infty} \frac{v'}{v} \leq \epsilon\eta^\diamond - \gamma_\diamond < 0 .$$

This implies that $v(t) \to 0$ as $t \to \infty$, contradicting our earlier result that the parasite persists uniformly strongly. So we have shown that there exists some $\epsilon > 0$ such that $\limsup_{t\to\infty} N(t) \geq \epsilon$ for every solution with $N(0) > 0$.

In order to replace $\limsup$ by $\liminf$ in the last statement, we apply Thm. A.32 (Thieme 2003) . We identify the state space of (12),

$$X = \left\{ (N, v_1, \ldots, v_n); N > 0, v_j \geq 0, \sum_{j=1}^{n} v_j \leq 1 \right\} ,$$

which we endow with the metric induced by the sum-norm. We choose the functional ρ on X as $\rho(x) = N$ for $x = (N, v_1, \ldots, v_n)$. In the language of chap. A.5 (Thieme 2003), we have just shown that the semiflow Φ generated on X by the solutions of (12) is uniformly weakly ρ-persistent. Notice that the compactness condition (C) in Thm. A.32 (Thieme 2003) holds as the set $\{\epsilon_1 \leq N \leq \epsilon_0\}$, $0 < \epsilon_1 < \epsilon_0$, is compact in X. It follows that the semiflow Φ is uniformly strongly ρ-persistent which translates into the host population being uniformly strongly persistent.

We consider the same type of selection functional as in sect. 6.2 (Ackleh and Allen 2003; 2005),

$$y_j = I_j^{\xi_j} I_1^{-\xi_1} ,$$

with ξ_j to be determined later. Then

$$\frac{y_j'}{y_j} = \xi_j \frac{I_j'}{I_j} - \xi_1 \frac{I_1'}{I_1} \,.$$

We substitute the appropriate differential equations from (9),

$$\frac{y_j'}{y_j} = \xi_j \eta_j \left(N - \sum_{k=1}^{n} I_k - \mathcal{R}_j^{-1} \right) - \xi_1 \eta_1 \left(N - \sum_{k=1}^{n} I_k - \mathcal{R}_1^{-1} \right) .$$

We choose $\xi_j = \eta_j^{-1}$. Then

$$\frac{y_j'}{y_j} = \mathcal{R}_1^{-1} - \mathcal{R}_j^{-1} \,.$$

We integrate this equation,

$$y_j(t) = y_j(0) \, \mathrm{e}^{-\delta_j t}, \quad \delta_j = \mathcal{R}_j^{-1} - \mathcal{R}_1^{-1} > 0 \,, \quad j \neq 1 \,.$$

We recall the definition of y_j,

$$I_j(t) = [y_j(0)]^{\eta_j} [I_1(t)]^{\eta_j / \eta_1} \, \mathrm{e}^{-\eta_j \delta_j t} \,. \tag{14}$$

The next result states that the strain with maximal replacement ratio does not go extinct.

Proposition 3. *Let $\mathcal{R}_1^{-1} > \mathcal{R}_j^{-1}$ for all $j = 2, \ldots, n$. Then the first strain uniformly strongly persists, i. e., there exists some $\epsilon > 0$ such that $\liminf_{t \to \infty} I_1(t) \geq \epsilon$ for all solutions with $I_1(0) > 0$.*

The proof has been banished into the appendix because of its length.

Theorem 2. *Let $\mathcal{R}_1^{-1} < \mathcal{R}_k^{-1}$ for all $k = 2, \ldots, n$. Let $j \in \{2, \ldots, n\}$ and $\eta_j \leq \eta_1$. Then $\frac{I_j(t)}{I_1(t)} \to 0$ as $t \to \infty$. In particular, $\frac{I_j(t)}{N(t)} \to 0$ as $t \to \infty$.*

Proof. By (14),

$$\frac{I_j(t)}{I_1(t)} = [y_j(0)]^{\eta_j} [I_1(t)]^{\frac{\eta_j}{\eta_1} - 1} \mathrm{e}^{-\eta_j \delta_j t} \,.$$

By Proposition 3, there exists some $\epsilon > 0$ such that $\liminf_{t \to \infty} I_1(t) \geq \epsilon$. Since $\frac{\eta_j}{\eta_1} \leq 1$,

$$\limsup_{t \to \infty} \frac{I_j(t)}{I_1(t)} \leq [y_j(0)]^{\eta_j} \epsilon^{\frac{\eta_j}{\eta_1} - 1} \limsup_{t \to \infty} \mathrm{e}^{-\delta_j \gamma_j t} = 0 \,.$$

The following result gives a condition under which the strain with maximal replacement ratio drives the suboptimal strains into extinction.

Theorem 3. *Let $\mu + \alpha_1 > q_1\beta$ and $\mathcal{R}_1^{-1} < \mathcal{R}_j^{-1}$ for all $j = 2,\ldots,n$. For $j = 2,\ldots,n$, let $\mu + \alpha_j > q_j\beta$ or $\eta_j \le \eta_1$. Then, for all solutions of (4) with $I_1(0) > 0$, the following hold: $I_j(t) \to 0$ for $j = 2,\ldots,n$, $\liminf_{t\to\infty} I_1(t) \ge \epsilon$ and $\limsup_{t\to\infty} N(t) \le c$ with $c \ge \epsilon > 0$ not depending on the initial data.*

Proof. Let J be the set of those $j \in \{1,\ldots,n\}$ such that $\mu + \alpha_j > q_j\beta$. By assumption, $1 \in J$. Consider a solution with $I_1(0) > 0$. Suppose that there exists some $c > 1$, which can be chosen arbitrarily large, and some $t_c > 0$ such that $N(t) > c \ge 1$ for all $t \ge t_c$. Let $j \notin J$. By assumption, $\eta_j \le \eta_1$. By (14), for $t \ge t_c$,

$$I_j(t) \le [y_j(0)]^{\eta_j}[N(t)]^{\eta_j/\eta_1}\, e^{-\eta_j\delta_j t} \le [y_j(0)]^{\eta_j} N(t)\, e^{-\eta_j\delta_j t}\,.$$

This implies

$$v_j(t) \le [y_j(0)]^{\eta_j}\, e^{-\eta_j\delta_j t}, \quad j \notin J, t \ge t_c\,.$$

Let $v_J(t) = \sum_{j\in J} v_j(t)$. Let $\eta_J = \min_{j\in J}\eta_j$ and $\nu = \max_{j\in J} a_j$. Then, for $t \ge t_c$, by (12),

$$v'_J \ge c\eta_J\left(1 - \sum_{j\notin J}[y_j(0)]^{\eta_j}\, e^{-\eta_j\delta_j t} - v_J\right)v_J - \nu v_J\,.$$

Let $\epsilon > 0$ which will be chosen sufficiently small later. By choosing $c > 1$ large enough, we can achieve that there exists some $t_\epsilon \ge t_c$ such that $v_J(t) \ge 1 - \epsilon$ for all $t \ge t_\epsilon$. By (12),

$$\frac{N'}{N} \le \beta - \mu - \sum_{k\in J} b_k v_k \le \beta - \mu - \left(\inf_{k\in J} b_k\right)v_J\,.$$

For $t \ge t_\epsilon$, recalling the definition of b_k in (6),

$$\frac{N'}{N} \le \beta - \mu - \inf_{k\in J}\left(\alpha_k + \beta(1 - q_k)\right)(1 - \epsilon)$$
$$= \max_{k\in J}\left[\beta(\epsilon + q_k(1 - \epsilon)) - \alpha_k(1 - \epsilon) - \mu\right]\,.$$

By assumption, the right hand side becomes negative, when $\epsilon > 0$ is chosen small enough, and $\frac{N'}{N} < 0$ for $t \ge t_\epsilon$. It follows that $N(t) \to 0$ as $t \to \infty$, This contradiction shows that there exists some $c > 0$ such that $\liminf_{t\to\infty} N(t) \le c$ for all solution with $I_1(0) > 0$. In order to have this result with $\limsup$ rather than $\liminf$, we are going to apply Thm. A.32 in Thieme (2003). To this end, we set

$$X = \left\{(N, I_1,\ldots,I_n) \in \mathbb{R}_+^{n+1}; I_1 > 0, \sum_{j=1}^n I_j \le N\right\}$$

and $\rho(x) = (1 + N)^{-1}$. In the language of sect. A.5 (Thieme 2003), we have shown that the semiflow induced by the solutions of (4) is uniformly weakly ρ-persistent. By Proposition 3, there exists some $\epsilon > 0$ such that $\liminf_{t\to\infty} I_1(t) \geq \epsilon$ for all solutions with $I_1(0) > 0$. Set $B = \{x \in X; I_1 \geq \epsilon/2\}$. Then condition (C) in sect. A.5 (Thieme 2003) is satisfied. By Thm. A.32 (Thieme 2003), the semiflow is uniformly strongly ρ-persistent, i. e., there exists some $c > 0$ such that $\limsup_{t\to\infty} N(t) \leq c$ for all solutions with $I_1(0) > 0$. In particular, $I_1 \leq N$ is bounded and the statement follows from (14).

While it is not very likely that the assumption $\eta_j \leq \eta_1$ is satisfied for all $j \geq 2$ or the assumption $\mu + \alpha_j > q_j\beta$ is satisfied for all j, the following example shows that the assumption in Theorem 3, $\mu + \alpha_j > q_j\beta$ or $\eta_j \leq \eta_1$ for all $j \geq 2$, can easily be satisfied.

Example 1. We consider a disease with direct transmission only (without the waterborne propagules we have considered in sect. 6.2.2) and without vertical transmission, $p_j = 0$. Let ζ again be the parasite replication rate. It is suggestive that the disease death rate α is an increasing functions of ζ and q, the fertility ratio of an infective to a susceptible individual, is a decreasing function of ζ. If γ does not depend on ζ or only weakly so, then the cumulative expected sojourn in the infectious stage $\frac{1}{\mu+\gamma+\alpha}$ will be either a decreasing function of ζ or a uni-modal function of ζ which increases on some interval $[0, \zeta_1]$ and decreases on $[0, \zeta_1]$ (see Fig. 23.13 (Anderson and May 1991) for myxomatosis; notice that, in this figure, $\mathcal{R}_0$ is proportional to the average sojourn time in the infectious stage). Since we only consider direct transmission, it is reasonable to assume that the transmission rate η is an increasing function of ζ or an uni-modal function of ζ which increases on an interval $[0, \zeta_2]$ and decreases on $[\zeta_2, \infty)$. Either way, it makes sense to assume that η grows more slowly than $\mu + \gamma + \alpha$ for high replication rates ζ such that $\mathcal{R} = \frac{\eta}{\mu+\gamma+\alpha}$ takes its maximum at some finite positive argument ζ^*. Since the demographic dynamics are typically much slower than the epidemic dynamics, β is close to μ such that $\mu + \alpha - \beta q > 0$ should be satisfied at $\zeta = \zeta^*$. For many diseases (obviously so for myxomatosis), it makes sense to assume that $0 \leq \zeta_1 \leq \zeta_2 \leq \infty$, i. e., the per capita transmission rate η takes its maximum at a higher replication rate than the cumulative expected sojourn in the infectious stage. This means that $\mathcal{R}$ is an increasing function on $[0, \zeta_1]$ and a decreasing function on $[\zeta_2, \infty)$ and so $\zeta_1 \leq \zeta^* \leq \zeta_2$. This implies that η increases on $[0, \zeta^*]$ and $\frac{1}{\mu+\gamma+\alpha}$ decreases on $[\zeta^*, \infty)$. Let now the first strain have the replication rate $\zeta_1 = \zeta^*$ such that $\mathcal{R}_1$ is the maximum of the relative replacement ratio as a function of ζ. As we argued above, $\mu + \alpha_1 - \beta q_1 > 0$. Any other strain $j > 1$ has a replication rate $\zeta_j \neq \zeta^* = \zeta_1$. If $\zeta_j < \zeta_1$, then $\eta_j < \eta_1$ because η increases on $[0, \zeta_1]$. If $\zeta_j > \zeta_1$, $\mu + \alpha_j - \beta q_j > \mu + \alpha_1 - \beta q_1 > 0$. This shows that the assumption of Theorem 3 are satisfied and its assertion holds.

6.5 Density-dependent per capita mortality and mass-action incidence

There are several ways in which Verhulst's celebrated logistic equation for the growth of a disease-free population,

$$N' = (\beta - \mu - \nu N)N \,,$$

can be adapted to a host-parasite system. Here νN is the density-dependent part of the per capita mortality $\mu(N) = \mu + \nu N$. If the disease did not interfere with the host dynamics except possibly adding disease related mortality, this would lead to a system

$$N' = (\beta - \mu - \nu N)N - \sum_{k=1}^{n} \alpha_k I_k \,,$$

$$I_j' = S\eta_j I_j - (\mu + \nu N + \alpha_j + \gamma_j)I_j \,, \quad j = 1,\ldots,n \,,$$

where $S = N - \sum_{k=1}^{n} I_k$ is the size of the susceptible part of the population. For such a model, coexistence of two different parasite strains is possible (Ackleh and Allen 2003; Andreasen and Pugliese 1995). As pointed out to me by Hans Metz, competitive exclusion still holds if infective individuals are too sick to reproduce and to take part in the competition for the resources the lack of which leads to the density-dependent part of the disease-independent mortality. Under this assumption the model takes the following form,

$$S' = (\beta - \mu - \nu S)S - S\sum_{k=1}^{n} \eta_k I_k \,,$$

$$I_j' = S\eta_j I_j - (\mu + \nu S + \alpha_j + \gamma_j)I_j \,, \quad j = 1,\ldots,n \,. \tag{15}$$

It is not difficult to see that the solutions of this system are bounded. The same analysis as in sect. 1.2 shows that all strains go extinct except those for which

$$\mathcal{S}_j = \frac{\eta_j - \nu}{\mu + \alpha_j + \gamma_j}$$

is maximal and positive.

Theorem 4. *If $\mathcal{S}_j < \mathcal{S}_1$ for $j = 2,\ldots,n$, then $I_j(t) \to 0$ as $t \to \infty$ for $t \to \infty$, $j = 2,\ldots,n$.*

While a maximization principle holds in this case, it is not a basic replacement ratio that is being maximized. If the number of susceptibles S is constant, the average duration of the infectious period (including death) is $\frac{1}{\mu+\nu S+\alpha_j+\gamma_j}$ and the respective replacement ratio is $\mathcal{R}_j(S) = \frac{\eta_j S}{\mu+\nu S+\alpha_j+\gamma_j}$. The basic replacement ratio is obtained by evaluating this expression at the carrying capacity

$S = K = \frac{\beta - \mu}{\nu}$ to which the population size converges in the absence of the disease, $\mathcal{R}_j(K) = \frac{\eta_j K}{\beta + \alpha_j + \gamma_j}$. It is not difficult to find examples where $\mathcal{R}_j(K)$ and $\mathcal{S}_j$ are maximized for different j. It should be mentioned that logistic growth is a rather particular case as the Verhulst equation is the special case of the Bernoulli equation $N' = (\beta - \mu - \nu N^\theta)N$, $\theta > 0$, for $\theta = 1$. If $\theta \neq 1$, a clear-cut maximization principle seems to hold no longer for the analog of the model discussed above.

6.6 Linear birth rates and standard incidence

Under standard incidence, depending on the situation, different strains can either coexist or exclude each other in a host population which would grow exponentially in the absence of the disease (Lipsitch and Nowak 1995a). I will show that low disease prevalence favors strains with higher net replacement ratio, while high disease prevalence favors strains with lower disease death rate. This also holds, when population growth would be limited by a nonlinear population death rate in absence of the disease. Standard incidence results, when the per capita contact rate $C(N)$ in a population of size N does not depend on N. We absorb the constant contact rate into the parameters η_j such that $C(N) = 1$. Now η_j is to be interpreted as per capita infection rate. System (4) specializes to

$$\left. \begin{array}{l} N' = (\beta - \mu(N))N - \sum_{k=1}^{n}(\alpha_k + \beta(1 - q_k))I_k \,, \\ I'_j = \beta q_j p_j I_j + \left(1 - \sum_{k=1}^{n} \frac{I_k}{N}\right)\eta_j I_j - \left(\mu(N) + \alpha_j + \gamma_j\right)I_j \,, \\ j = 1, \ldots, n \,. \end{array} \right\} \qquad (16)$$

In terms of the proportions of infectives with strain j, system (5) specializes to

$$\left. \begin{array}{l} \frac{N'}{N} = \beta - \mu(N) - \sum_{k=1}^{n} b_k v_k \,, \\ \frac{v'_j}{v_j} = \left(1 - \sum_{k=1}^{n} v_k\right)\eta_j - a_j + \sum_{k=1}^{n} b_k v_k \,, \quad a_j > b_j \,, \end{array} \right\} \qquad (17)$$

and (7) specializes to

$$v' = (1 - v)\sum_{j=1}^{n} \eta_j v_j - \sum_{j=1}^{n} a_j v_j + v\sum_{j=1}^{m} b_j v_j \,. \qquad (18)$$

Recall the definitions of a_j and b_j in (6). If the host population grows exponentially in the absence of the disease (μ does not depend on N and is smaller that β), system (16) is a *homogeneous* differential equation (Hadeler 1991; Hadeler 1992). Homogeneous differential equations have no stationary solutions (steady states), except for rare combinations of parameters; the role of stationary solutions are taken by exponential solutions (Hadeler 1991; Hadeler 1992). This means, at 'equilibrium', in the presence of the disease, the host population continues to show exponential increase (at the same or a lower rate than in the absence of the disease) or is converted to exponential decline (Busenberg and Hadeler 1990; Busenberg et al. 1991; Busenberg and Cooke 1993; Hethcote et al. 1996; Thieme 1992). Either way, the proportions v_j of strain j infectives become much more relevant than the absolute numbers I_j. For this epidemiologic reason, we will now study strain competition in terms of proportions. There is also a mathematical benefit, as the equations for the proportions are decoupled from the equation for the host population and we have reduced our system by one dimension which makes its analysis much easier. Another nice feature is the elimination of the natural per capita mortality rate from the equations for the proportions which makes the following considerations largely independent of as to whether the natural population death rates are linear or nonlinear.

For a moment, let us consider the situation where there is just one strain,

$$\frac{v_1'}{v_1} = \left(1 - v_1\right)\eta_1 - a_1 + b_1 v_1 .$$

Assume $a_1 > b_1$. It is easy to see that the disease dies out in proportion ($v_1(t) \to 0$ as $t \to \infty$) if $\frac{\eta_1}{a_1} \leq 1$, and persists in proportion ($\liminf_{t \to \infty} v_1(t) > 0$, actually $v_1(t)$ converges to a positive limit) if $\frac{\eta_1}{a_1} > 1$. The number

$$\mathcal{R}_1^\circ = \frac{\eta_1}{a_1} = \frac{\eta_1}{\alpha_1 + \gamma_1 + \beta(1 - q_1 p_1)}$$

has the form of a basic replacement ratio, except that the per capita mortality rate μ has been replaced by $\beta(1 - q_1 p_1)$. The term 'basic' is appropriate here because, due to the choice of standard incidence, η_1 is the transmission rate of an average infective individual in an otherwise susceptible population. If $\beta > \mu$ and there is no vertical transmission, $p_1 = 0$, the replacement of μ by β can be interpreted as the proportional effect of infections being discounted by the growth of the population. The discount effect can be weakened or even turned upside down, if there is vertical transmission and no or only weak fertility reduction by the disease. Partially adopting the language of sect. 2.13.2 (Busenberg and Cooke 1993), we call $\mathcal{R}_1^\circ$ the *net replacement ratio* of strain 1. Analogously, we call

$$\mathcal{R}_j^\circ = \frac{\eta_j}{a_j} = \frac{\eta_j}{\alpha_j + \gamma_j + \beta(1 - q_j p_j)} \tag{19}$$

the *net replacement ratio* of strain j.

In order to find out what strain competition selects for, we fix j and again use the functional

$$z = v_n^{\xi_n} v_j^{-\xi_j} , \tag{20}$$

with $\xi_j > 0$ to be determined later (Ackleh and Allen 2003; 2005). Then

$$\frac{z'}{z} = \xi_n \frac{v_n'}{v_n} - \xi_j \frac{v_j'}{v_j} .$$

We substitute the differential equations (17),

$$\frac{z'}{z} = \xi_n\Big((1-v)\eta_n - a_n + \sum_{k=1}^{n} b_k v_k\Big)$$
$$- \xi_j\Big((1-v)\eta_j - a_j + \sum_{k=1}^{n} b_k v_k\Big) . \tag{21}$$

6.6.1 Selection for high net replacement ratio

By analogy, our previous results suggest that competition selects for strains with maximal net replacement ratio. This is the case, indeed, under the proviso that disease prevalence is low. We choose $\xi_j = \frac{1}{\eta_j}$ in (20). By (21),

$$\frac{z'(t)}{z(t)} = \frac{1}{\mathcal{R}_j^\circ} - \frac{1}{\mathcal{R}_n^\circ} + \Big(\frac{1}{\eta_n} - \frac{1}{\eta_j}\Big) \sum_{k=1}^{n} b_k v_k .$$

Proposition 4. *Assume that, for some* $j < n$, $\mathcal{R}_n^\circ < \mathcal{R}_j^\circ$ *and*

$$\frac{1}{\mathcal{R}_j^\circ} - \frac{1}{\mathcal{R}_n^\circ} + \Big(\frac{1}{\eta_n} - \frac{1}{\eta_j}\Big) \limsup_{t\to\infty}\Big(\sum_{k=1}^{n} b_k v_k\Big)^\infty < 0 .$$

Then $\frac{I_n(t)}{N(t)} \to 0$ *as* $t \to \infty$.

If disease prevalence is low for large times (all v_k are small), differences in net replacement ratio matter much more than differences in η_j for the assumption in this proposition.

Proof. First let $\eta_n \geq \eta_j$. Then $\sup_{t\geq 0} \frac{z'(t)}{z(t)} < 0$. If $\eta_n < \eta_j$,

$$\limsup_{t\to\infty} \frac{z'(t)}{z(t)} \leq \frac{1}{\mathcal{R}_j^\circ} - \frac{1}{\mathcal{R}_n^\circ} + \Big(\frac{1}{\eta_n} - \frac{1}{\eta_j}\Big) \limsup_{t\to\infty}\Big(\sum_{k=1}^{n} b_k v_k\Big)^\infty < 0$$

by assumption. In either case, $z(t) \to 0$ as $t \to \infty$ and $v_n(t) = \frac{I_n(t)}{N(t)} \to 0$ by (20).

Set $v^\infty = \limsup_{t\to\infty} v(t)$.

Proposition 5. *Assume that there is no fertility reduction in infectives,* $q_j = 1$, *and no vertical transmission,* $p_j = 0$, *for all* j. *Then* $v^\infty \leq \Big[1 - \frac{\beta}{\bar\zeta}\Big]_+$ *where* $\bar\zeta = \max_{j=1}^{m}(\eta_j - \alpha_j)$.

Proof. Under these assumptions, by (6) and (18),

$$v' \leq (1-v) \sum_{k=1}^{n} (\eta_k - \alpha_k) v_k - \beta v \leq (1-v)\bar\zeta v - \beta v .$$

If the per capita birth rate is sufficiently large, competition essentially selects for the largest net replacement ratio.

Theorem 5. *Assume that there is no fertility reduction in infectives, $q_j = 1$ and no vertical transmission, $p_j = 0$, for all j. Assume that, for some $j < n$, $\mathcal{R}_n^\circ < \mathcal{R}_j^\circ$ and*

$$\frac{1}{\mathcal{R}_j^\circ} - \frac{1}{\mathcal{R}_n^\circ} + \left(\frac{1}{\eta_n} - \frac{1}{\eta_j}\right)\bar{\alpha}\left[1 - \frac{\beta}{\bar{\zeta}}\right]_+ < 0\,,$$

where $\bar{\alpha} = \max_{k=1}^n \alpha_k$. Then $\frac{I_n(t)}{N(t)} \to 0$ as $t \to \infty$.

6.6.2 Selection for low disease fatality under high disease prevalence

In order to see what competition selects for if disease prevalence is high, we choose $\xi_j = 1 = \xi_n$ in (20). High disease prevalence ($> 80\%$) has been observed in the Arizona grass *Festucca arizonica* for the endophytic fungus *Neotyphodium* (Schulthess and Faeth 1998). By (21),

$$\frac{z'}{z} = (1 - v)(\eta_n - \eta_j) + a_j - a_n\,.$$

We define $v_\infty = \liminf_{t\to\infty} v(t)$.

Proposition 6. *Let there exist some $j < n$ such that $a_j < a_n$ and $a_j - a_n + (1 - v_\infty)(\eta_n - \eta_j) < 0$. Then $\frac{I_n(t)}{N(t)} \to 0$ as $t \to \infty$.*

If disease prevalence is high for large times, i. e., v_∞ is close to 1, differences in a_j matter much for than differences in η_j for the assumption in this proposition.

Proof. First assume that $\eta_n \leq \eta_j$. Then $\frac{z'}{z} \leq a_j - a_n < 0$. Now assume that $\eta_n > \eta_j$. Then

$$\limsup_{t\to\infty} \frac{z'}{z} \leq a_j - a_n + (1 - v_\infty)(\eta_n - \eta_j) < 0\,.$$

In either case, $z(t) \to 0$ as $t \to \infty$ and so $v_n(t) \leq z(t)^{1/\xi_n} \to 0$ as $t \to \infty$.

We show that the disease prevalence is high if there is no recovery from the disease and both horizontal and vertical transmission are strong. This has been numerically demonstrated for mass action incidence (Lipsitch et al. 1995b); in the mathematically easier case of standard incidence, it is possible to provide qualitative estimates.

Proposition 7. *Let $\mathcal{R}_j^{\circ} > 1$ and $\gamma_j = 0$ for $j = 1, \ldots, n$. Then*

$$1 - v_{\infty} \leq \beta \frac{\psi^*}{\phi_* + \beta\psi^*} \,,$$

where

$$\phi_* = \min_{j=1}^{n} \left(\eta_j - \alpha_j - \beta(1 - q_j p_j)\right) > 0$$

and

$$\psi^* = \max_{j=1}^{n} q_j(1 - p_j) \,.$$

Proof. By (7),

$$v' = (1 - v) \sum_{j=1}^{n} \eta_j v_j - \sum_{j=1}^{n} a_j v_j + v \sum_{j=1}^{n} b_j v_j \,.$$

By (6), since $\gamma_j = 0$,

$$v' = (1 - v) \sum_{j=1}^{n} \left(\eta_j - \alpha_j - \beta(1 - q_j p_j)\right) v_j - v\beta \sum_{j=1}^{n} q_j(1 - p_j) v_j \,.$$

Since $\mathcal{R}_j^{\circ} > 1$ for all j, $\phi_* > 0$ and

$$v' \geq \phi_*(1 - v)v - \beta\psi^* v^2 \,.$$

This implies that $v_{\infty} \geq \frac{\phi_*}{\phi_* + \beta\psi^*}$ and the assertion follows.

If both horizontal and vertical transmission are high for all strains, competition essentially selects for some combination of the smallest disease fatality, the highest vertical transmission rate and the lowest fertility reduction. The latter two may conflict with each other, as the vertical infection of offspring may go along with a decrease of its vitality.

Proposition 8. *Let $\gamma_j = 0$ and $\mathcal{R}_j^{\circ} > 1$ for $j = 1, \ldots, n$. Assume that there exists some $j \in \{2, \ldots, n\}$ such that $\alpha_j < \alpha_n$ and*

$$\alpha_j - \alpha_n + \beta\left(\frac{(\eta_n - \eta_j)\psi^*}{\phi_* + \beta\psi^*} + q_n p_n - q_j p_j\right) < 0 \,,$$

with ϕ_* and ψ^* as in Proposition 7. Then $\frac{I_n(t)}{N(t)} \to 0$ as $t \to \infty$.

We can rewrite the condition in Proposition 8 as

$$\alpha_j + \beta(1 - q_j p_j) - \left(\alpha_n + \beta(1 - q_n p_n)\right) + (\eta_n - \eta_j)\frac{\beta\psi^*}{\phi_* + \beta\psi^*} < 0 \,,$$

which shows that, for the formula (19) for the net replacement ratio, strain competition selects much more strongly for a small denominator than for a large numerator.

We must caution, though, that this result holds under the assumption that no one recovers from the disease and that all competing strains have a net replacement ratio above 1. If recovery is incorporated into the model, it becomes much harder to estimate the disease prevalence, and typically two types of conditions rather than one need to be made for one strain to outcompete another (Ackleh and Allen 2005).

6.6.3 Persistence of protective subthreshold strains

A subthreshold strain (whose net replacement ratio is smaller than one) dies out, if it is the only strain circulating in the population. Can it possible persist if it protects against a more virulent superthreshold strain? This surprising phenomenon has been observed in numerical simulations for mass action incidence (Lipsitch et al. 1996); in the mathematically easier case of standard incidence, it is possible to give precise qualitative conditions for its occurrence.

By (17), at most two strains can coexist at equilibrium, except for exceptional values which form a set of measure 0. While this does not necessarily imply that all but two strains die out in proportion, we restrict the forthcoming analysis to two strains.

One-strain equilibria

We assume that the first strain, which is a superthreshold strain, is at equilibrium with no other strain being present,

$$v_1^\diamond = \frac{\eta_1 - a_1}{\eta_1 - b_1}, \quad \eta_1 > a_1 > b_1,$$
$$v_2^\diamond = 0.$$

(22)

At this equilibrium, the equation for the second strain becomes

$$\frac{v_2'}{v_2} = \left(1 - v_1^\diamond\right)\eta_2 - a_2 + b_1 v_1^\diamond > 0.$$

This shows that the second strain (sub- or super-threshold) can invade if and only if

$$\eta_2 - a_2 > v_1^\diamond(\eta_2 - b_1).$$

In general, an invading strain does not necessarily persist (Mylius and Diekmann 2001); but since we have a planar system and no multiple boundary equilibria, it follows from standard persistence theory as we have used it earlier that the second strain is uniformly strongly persistent. If the second strain dies out when left on its own, then both strains coexist.

Theorem 6. *Let the number of circulating parasite strains be $n = 2$. If $\eta_2 - a_2 > v_1^\diamond(\eta_2 - b_1) > 0$, the second strain is strongly uniformly persistent: there exists some $\epsilon > 0$ such that $\liminf_{t\to\infty} v_2(t) > 0$ whenever $v_2(0) > 0$, with the eventual lower bound ϵ being independent of the initial data. If $\mathcal{R}_2^\circ < 1 < \mathcal{R}_1^\circ$, then both strains are strongly uniformly persistent in this sense.*

In order to illustrate that the conditions of Theorem 6 are feasible we consider the extreme case that the second strain transmits vertically only, i. e., $\eta_2 = 0 = \mathcal{R}_2^\circ$. The condition above takes the form

$$b_1 v_1^\diamond > a_2 \,,$$

which implies that $a_1 > b_1 > a_2 \geq b_2$ (because $v_1^\diamond \in (0,1)$) and is equivalent to

$$\frac{a_2}{b_1} < \frac{\eta_1 - a_1}{\eta_1 - b_1} \,.$$

We recall the definitions of a_j an b_j and find the following condition for the uniform strong persistence of both strains,

$$\frac{\alpha_2 + \gamma_2 + \beta(1 - q_2 p_2)}{\alpha_1 + \beta(1 - q_1)} < \frac{\eta_1 - (\alpha_1 + \gamma_1 + \beta(1 - q_1 p_1))}{\eta_1 - (\alpha_1 + \beta(1 - q_1))} \,,$$

with both numerator and denominator of the second fraction being positive. The left hand side of this inequality can be brought close to 0, if there is no recovery from strain 2, $\gamma_2 = 0$, the vertical transmission of the second strain is almost perfect, p_2 close to 1, and the virulence of the second strain is sufficiently less than that of the first strain, in particular $\alpha_2 \ll \alpha_1$ and $q_2 \ll q_1$. The right hand side of the inequality can be brought close to 1, if at least one of the following holds:

- there is no recovery from the first strain as well, $\gamma_1 = 0$, and the vertical transmission of the first strain is also almost perfect,

or

- the per capita transmission rate of strain 1, η_1, is large enough.

6.7 Discussion

Emerging and re-emerging infectious diseases (Garrett 1995; Ewald and De Leo 2002) have led to a renewed interest in host-parasite systems, and their mathematical modeling (Brauer and Castillo-Chavez 2001; Brauer and van den Driessche 2002; Dieckmann et al. 2002b; Castillo-Chavez 2002a, 2002b; Diekmann and Heesterbeek 2000; Hethcote 2000; Rass and Radcliffe 2003; Thieme 2003) (see the bibliographic remarks in Chap. 17 (Thieme 2003) for more references). The fact that not only humans and their food sources (domestic animals and agronomic plants), but also natural animal and plant populations are afflicted, has also directed attention to the fascinating role of parasites in ecosystems (Grenfell and Dobson 1995; Hudson et al. 2002; O'Neill et al. 1997; Hatcher et al. 1999).

In this paper, we revisit the question of competition between several parasite strains for one host and the evolution of virulence. In particular, we investigate the validity of the often made hypothesis that the strain with maximal replacement ratio outcompetes the other strains. This hypothesis has been the starting point for investigating whether evolution would lead to low, intermediate, or high virulence of the parasite. From a host perspective (which we adopt here), virulence is the degree by which the parasite lowers the basic reproduction ratio of the host by increasing the host mortality and morbidity and/or reducing the host fertility.

For mathematical managability, we assume that infection with one strain infers complete protection against other strains during the infection and complete immunity and cross-immunity during the recovery period (if there is one). We first extend the known result that the principle of maximizing the basic replacement ratio, $\mathcal{R}^\circ$, holds for host populations which, in the absence of the disease, are exclusively regulated by a nonlinear birth rate. In view of the discussion as to whether mass action, standard or some interpolating form of incidence are appropriate (see sect. 2.1 (Hethcote 2000) for a discussion and references) and the big differences that they can make for the dynamics of the disease (Gao et al. 1995), we emphasize that this result holds for all forms of incidence provided that transmission is only horizontal. Its proof does no longer work when a latent period or the release of long-living propagules (Ewald and De Leo 2002; Day 2002a) is added to the model. For short-living propagules the result can be salvaged by a quasi-steady state approach. The shifts in virulence evolution produced by a change of hygiene in diseases with both direct and propagule transmission may be even more dramatic as previously thought (Ewald and De Leo 2002 and Day 2002a).

If vertical transmission is added (Lipsitch et al. 1996) to the model, coexistence of different strains becomes possible, and strains with lower $\mathcal{R}^\circ$ can outcompete strains with higher $\mathcal{R}^\circ$. The simulations which show this result have been performed for mass action incidence (Lipsitch et al. 1996), but presumably coexistence may occur for standard incidence as well (see below).

The principle of $\mathcal{R}^\circ$ maximization still holds if both the population birth rate and the population death rate are linear, but the exponential growth rate of the host population (without the disease) is low enough that the disease can limit population growth. Further the incidence is assumed to be of mass action type and immunity nonexistent. If the concept of $\mathcal{R}^\circ$ is appropriately extended, vertical transmission can be included.

If host population growth is limited by a nonlinear mortality rate, coexistence of different strains is possible (Ackleh and Allen 2003; Andreasen and Pugliese 1995) under mass action incidence, but it not clear whether strains with low $\mathcal{R}^\circ$ can outcompete strains with higher $\mathcal{R}^\circ$. We show that both phenomena are possible if standard incidence is assumed and $\mathcal{R}^\circ$ is interpreted as net replacement ratio: if both horizontal and vertical transmission are high for all circulating strains (in particular if they all have $\mathcal{R}^\circ > 1$) and if there is no disease recovery (resulting in high disease prevalence), competition selects for a combination of low disease fatality, high vertical transmissibility, and low fertility reduction (with the latter two possibly conflicting with each other) rather than for a large net replacement ratio.

If the per capita birth rate, β, is large enough to cause low disease prevalence (the disease can hardly keep pace with a fast population turnover), $\mathcal{R}^\circ$ maximization holds cum grano salis, i. e., strains are selected which are close to maximal net replacement ratio. Otherwise, coexistence of different strains can occur, even to the degree that a strain with exclusive but incomplete vertical transmission which would go extinct on its own can persist by protecting against a more virulent strain which also transmits horizontally. More generally, a subthreshold strain with little virulence can coexist with a highly virulent superthreshold strain by providing protection against it. This surprising phenomenon has already been observed in simulations for mass action incidence (Lipsitch et al. 1996), but in the mathematically easier case of standard incidence it is possible to give precise conditions under which it occurs. (Very recent work shows that this is possible for general incidence (Faeth et al., to appear)). A vertically transmitted parasite with no apparent horizontal transmission is the endophytic fungus *Neotyphodium* in the Arizona grass *Festucca arizonica*. So far, the persistence of this endophyte is a mystery, because its vertical transmission is not perfect and no competitive advantages have been found that it might give to infected plants (Faeth and Bultman 2002; Faeth 2002; Faeth and Sullivan 2003). Protection against a more virulent strain or a more virulent other parasite would be a possible explanation, but a candidate has been elusive so far. Persistence of subthreshold strains (whose basic replacement ratio is smaller than one) is also possible by an mechanism opposite to protection where the subthreshold strain can infect individuals which have recovered from infection by a superthreshold strain (Nuño et al. to appear).

6.8 Appendix

We prove Proposition 3: *Let $\mathcal{R}_1^{-1} < \mathcal{R}_j^{-1}$ for all $j = 2,\ldots,n$. Then the first strain uniformly strongly persists, i. e., there exists some $\epsilon > 0$ such that $\liminf_{t\to\infty} I_1(t) \geq \epsilon$ for all solutions with $I_1(0) > 0$.*

Proof. By Proposition 2, there exists some $\epsilon_0 > 0$ such that

$$\liminf_{t\to\infty} N(t) > \epsilon_0$$

148   Horst R. Thieme

for all solutions of (9) with $N(0) > 0$. Since $\beta > \mu$ by assumption, we can choose $\epsilon > 0$ small enough that

$$(\beta - \mu)\epsilon_0 - (\alpha_1 + \beta(1 - q_1))\epsilon > 0 .$$

Assume that

$$\limsup_{t \to \infty} I_1(t) < \epsilon .$$

By (14), $I_j(t) \to 0$ as $t \to \infty$ for $j = 2, \ldots, n$. By the first equation in (9), $\liminf_{t \to \infty} N'(t) > 0$. This implies $N(t) \to \infty$ as $t \to \infty$. By the equation for I_1 in (9), $\frac{I_1'(t)}{I_1(t)} \to \infty$ as $t \to \infty$ and so $I_1(t) \to \infty$ as $t \to \infty$. This contradiction shows that $\limsup_{t \to \infty} I_1(t) \geq \epsilon$ for every solution with $I_1(0) > 0$. We apply Thm. 6.2 (Thieme 1993). We identify the state space

$$X = \left\{ (N, I_1, \ldots, I_n) \in \mathbb{R}_+^{n+1}, I_j \geq 0, \sum_{j=1}^{n} I_j \leq N \right\}$$

and the semiflow Φ induced by the solutions of (9) on X. We set $X_1 = \{(N, I_1, \ldots, I_n) \in X; I_1 > 0\}$ and $X_2 = \{(N, I_1, \ldots, I_n) \in X; I_1 = 0\}$. Then X is the disjoint union of X_1 and X_2, X_2 is closed in X and X_1 is open in X, and Φ is a continuous semiflow on X_1. By (14), $\frac{I_j(t)}{[I_1(t)]^{\eta_j/\eta_1}} \to 0$ as $t \to \infty$. To satisfy Ass. 6.1 (Thieme 1993) we choose

$$Y_1 = \{(N, I_1, \ldots, I_n) \in X_1; \, I_j \leq I_1^{\eta_j/\eta_1}, j = 2, \ldots, n\} .$$

Then part (A) is satisfied. Parts $(C_{6.1})$, and $(C_{6.2})$ are trivially satisfied because every bounded set in $\mathbb{R}^{n+1}$ has compact closure. The following is a stronger version of part (R) of the Assumptions:

$(\tilde{R})$ For any sufficiently $\epsilon > 0$ there are a bounded subset D of X_1 and some $\delta \in (0, \epsilon)$ and time $t > 0$ with the following properties:
  (i)   There is no element $x \in Y_1 \setminus D$ such that $d(x, X_2) = \epsilon$ and $d(\Phi_s(x), X_2) < \epsilon$ for all $s \in (0, t]$.
  (ii) If $x \in Y_1 \setminus D$ and $r \in (0, t]$ is such that $d(x, X_2) = \epsilon = d(\Phi_r(x), X_2)$ and $d(\Phi_s(x), X_2) < \epsilon$ for all $0 < s < r$, then $d(\Phi_s(x), X_2) \geq \delta$ for all $0 \leq s \leq r$.
  (iii) $D \cap Y_1$ is bounded.

Notice that $d(x, X_2) = I_1$ for $x = (N, I_1, \ldots, I_n)$. Let $\epsilon \in (0, 1]$. Let $t = 1$ and $\delta = \epsilon/2$. We choose $N_\epsilon > 0$ sufficiently large that

$$(\beta - \mu)N_\epsilon - (\alpha_1 + \beta(1 - q_1))\epsilon - \sum_{j=2}^{n} (\alpha_j + \beta(1 - q_j))\epsilon^{\eta_j/\eta_1} > 0$$

and

$$N_\epsilon - \epsilon - \sum_{j=2}^{n} \epsilon^{\eta_j/\eta_1} - \sum_{j=1}^{n} \frac{1}{\mathcal{R}_j^\circ} > 0$$

and $D = \{(N, I_1, \ldots, I_n) \in X_1; N \leq N_\epsilon\}$. Then D is bounded and (iii) is satisfied. Suppose that $x \in Y_1 \backslash D$ such that $d(x, X_2) = \epsilon$. Let $(N(s), I_1(s), \ldots, I_n(s)) = \Phi_s(x)$. Then $N(0) > N_\epsilon$, $I_1(0) = \epsilon$ and $I_j(0) \leq \epsilon^{\eta_j/\eta_1}$ for $j = 2, \ldots, n$. By choice of N_ϵ and the differential equation for I_1 in (9),, $I_1'(0) > 0$ and $d(\Phi_s(x), X_2) = I_1(s) > I_1(0) = \epsilon$ for some $s \in [0, t]$. So (i) is satisfied.

Assume that $x \in Y_1 \setminus D$ and that $\Phi_s(x) \in Y_1$ for all $s \geq 0$. Again this implies $N(0) > N_\epsilon$. Further let $0 < r$ such that $d(x, X_2) = \epsilon = d(\Phi_r(x), X_2)$ and $d(x, \Phi_s(x)) \leq \epsilon$ for all $s \in [0, r]$. Then $I_1(s) \leq \epsilon$ for all $s \in [0, r]$ and $I_j(s) \leq \epsilon^{\eta_j/\eta_1}$. By (9) and our choice of N_ϵ and by $N(0) > N_\epsilon$, $N(s) \geq N_\epsilon$ for all $s \in [0, r]$ and $I_1(s) \geq I_1(0) = \epsilon$. So part (ii) is trivially satisfied. By Thm. 6.2 (Thieme 1993), X_2 is a uniform strong repeller for X_1. By our choice of X_1 and X_2, this translates into the strong persistence of the first strain.

Acknowledgement. The author thanks Stan Faeth and an anonymous referee for useful comments and Hans Metz for the model in sect. 6.5.

References

1. Ackleh, A.S. and L.J.S. Allen (2003), 153-168. Competitive exlusion and coexistence for pathogens in an epidemic model with variable population size, *J. Math. Biol.*, **47**, 153–168.
2. Ackleh, A.S. and L.J.S. Allen, Competitive exclusion in SIS and SIR epidemic models with total cross immunity and density-dependent host mortality, *Discrete and Continuous Dynamical Systems Series B* **5**(2005), 175–188.
3. Anderson, R.M. and R.M. May (1982), *Population Biology of Infectious Diseases*, Springer, Dahlem Konferenzen, Berlin.
4. Anderson, R.M. and R.M. May (1991), *Infectious Diseases of Humans*, Oxford University Press, Oxford.
5. Andreasen, V. (1997), The dynamics of cocirculating influenza strains conferring partial cross-immunity, *J. Math. Biol.*, **35**, 825–842.
6. Andreasen, V. and A. Pugliese (1995), Pathogen coexistence induced by density dependent host mortality, *J. Theor. Biol.*, **177**, 159–165.
7. Andreasen, V., J. Lin and S.A. Levin (1997), The dynamics of co-circulating influenza strains conferring partial cross-immunity, *J. Math. Biol.*, **35**, 825–842.
8. Bowers, R.G. and J. Turner (1997), Community structure and the interplay between interspecific infection and competition, *J. Theor. Biol.*, **187**, 95–109.
9. Brauer, F. and C. Castillo-Chavez (2001), *Mathematical Models in Population Biology and Epidemiology*, Springer, New York.

10. Brauer, F. and P. van den Driessche (2002), Some directions for mathematical epidemiology, In S. Ruan, G. Wolkowicz and J. Wu, *Dynamical Systems and their Applications to Biology*, 95 E12, Fields Institute Communications, American Mathematical Society, Providence, RI.
11. Bremermann, H.J. and H.R. Thieme (1989), A competitive exclusion principle for pathogen virulence, *J. Math. Biology*, **27**, 179–190.
12. Brunner, J. (2004), *Ecology of an Amphibian Pathogen: Transmission, Persistence, and Virulence*, Dissertation, Arizona State University, Tempe.
13. Bull, J.J., I.J. Molineux and W.R. Rice (1991), Selection of benevolence in a host-parasite system, *Evolution*, **45**, 875–882.
14. Busenberg, S., K.L. Cooke and M.A. Pozio (1983), Analysis of a model of a vertically transmitted disease, *J. Math. Biol.*, **17**, 305–329.
15. Busenberg, S.N. and K.L. Cooke (1993), *Vertically Transmitted Diseases*, Springer, Berlin Heidelberg.
16. Busenberg, S.N. and K.P. Hadeler (1990), Demography and Epidemics, *Math. Biosci.*, **101**, 63–74
17. Busenberg, S.N., K.L. Cooke and H.R. Thieme (1991), Demographic change and persistence of HIV/AIDS in a heterogeneous population, *SIAM J. Appl. Math.*, **51**, 1030–1052.
18. Castillo-Chavez, C. and J.X. Velasco-Hernández (1998), On the relationship between evolution of virulence and host demography, *J. Theor. Biol.*, **192**, 437–444.
19. Castillo-Chavez, C., S. Blower, P. van den Driessche, D. Kirschner and A.-A. Yakubu (2002a), *Mathematical Approaches for Emerging and Reemerging Infectious Diseases: an Introduction*, Springer, New York.
20. Castillo-Chavez, C., S. Blower, P. van den Driessche, D. Kirschner and A.-A. Yakubu (2002b), *Mathematical Approaches for Emerging and Reemerging Infectious Diseases: Models, Methods and Theory*, Springer, New York.
21. Castillo-Chavez, C., W. Huang and J. Li (1996), Competitive exclusion in gonorrhea models and other sexually transmitted diseases, *SIAM J. Appl. Math.*, **56**, 494–508.
22. Castillo-Chavez, C., W. Huang and J. Li (1999), Competitive exclusion and coexistence of multiple strains in an SIS STD model, *SIAM J. Appl. Math.*, **59**, 1790–1811.
23. Castillo-Chavez, C. and H.R. Thieme (1995), Asymptotically autonomous epidemic models, *Mathematical Population Dynamics: Analysis of Heterogeneity* Vol. One: Theory of Epidemics (O. Arino, D. Axelrod, M. Kimmel, M. Langlais; eds.), 33–50. Wuerz, Winnipeg.
24. Charles, S., S. Morand, J.L. Chassé and P. Auger (2002), Host patch selection induced by parasitism: basic reproduction ratio R_0 and optimal virulence, *Theor. Pop. Biol.*, **62**, 97–109.
25. Clayton, D.H.; J. Moore (1997), *Host-Parasite Evolution: General Principles and Avian Models*, Oxford University Press, Oxford.
26. Davies, C.M.; J.P. Webster, M.E.J. Woolhouse (2000), Trade-offs in the evolution of virulence in an indirectly transmitted macroparasite, *Proc. Royal Soc. London, B*, **268**, 251–257.
27. Dawns, J.H.P., J.R. Gog (2002), The onset of oscillatory dynamics of models of multiple disease strains, *J. Math. Biol.*, **45**, (2002), 471–510.
28. Day, T. (2001), Parasite transmission modes and the evolution of virulence, *Evolution*, **55**, 2389–2400.

29. Day, T. (2002a), Virulence evolution via host exploitation and toxin production in spore-producing pathogens, *Ecology Letters*, **5**, 471–476.
30. Day, T. (2002b), On the evolution of virulence and the relationship between various measures of mortality, *Proc. Royal Soc. London, B* **269**, 1317–1323.
31. Day, T. and S.R. Proulx (2004), A general theory for the evolutionary dynamics of virulence, *Amer. Nat.*, **163**, E40–E63.
32. Dieckmann, U. (2002a), Adaptive dynamics of pathogen-host interactions, In Dieckmann, U., J.A.J. Metz, M.W. Sabelis, K. Sigmund, *Adaptive Dynamics of Infectious Diseases: in Pursuit of Virulence Management*, International Institute for Applied Systems Analysis, Cambridge University Press, Cambridge. pp. 39–59
33. Dieckmann, U., J.A.J. Metz, M.W. Sabelis, K. Sigmund (2002b), *Adaptive Dynamics of Infectious Diseases: in Pursuit of Virulence Management*, International Institute for Applied Systems Analysis, Cambridge University Press, Cambridge.
34. Diekmann, O. and J.A.P. Heesterbeek (2000), *Mathematical Epidemiology of Infectious Diseases: Model Building, Analysis and Interpretation*, Wiley, Chichester.
35. Dietz, K. (1975), Transmission and control of arbovirus diseases, *Epidemiology* (D. Ludwig, K.L. Cooke, eds.), 104–119, SIAM, Philadelphia.
36. Esteva, L. and C. Vargas (2003), Coexistence of different serotypes of dengue virus, *J. Math. Biol.*, **46**, 31–47.
37. Ewald, P.W. (1984), *The Evolution of Infectious Disease*, Oxford University Press, Oxford.
38. Ewald, P.W. (2002), *Plague Time: the New Germ Theory of Disease*, Anchor Books, New York.
39. Ewald, P.W. and G. De Leo (2002), Alternative transmission modes and the evolution of virulence, In Dieckmann, U.; J.A.J. Metz, M.W. Sabelis, K. Sigmund, *Adaptive Dynamics of Infectious Diseases: in Pursuit of Virulence Management*, International Institute for Applied Systems Analysis, Cambridge University Press, Cambridge. pp. 10-25
40. Faeth, S.H. (2002), Are endophytic fungi defensive plant mutualists?, *Oikos*, **98**, 25–36.
41. Faeth, S.H. and T.L. Bultman (2002), Endophytic fungi and interactions among host plants, herbivores, and natural enemies, In Tscharntke, T. and B.A. Hawkins, *Multitrophic Level Interactions*, 89–123, Cambridge University Press, Cambridge.
42. Faeth, S.H. and T.J. Sullivan (2003), Mutualistic asexual endophytes in a native grass are usually parasitic, *The American Naturalist*, **161**, 310–325.
43. Feath, S.H., K.P. Hadeler, H.R. Thieme, An apparent paradox of horizontal and vertical desease transmission, *J. Biol. Dyn.* (to appear)
44. Feng, Z. and J.X. Velasco-Hernández (1997), Competitive exclusion in a vector-host model for dengue fever, *J. Math. Biol.*, **35**, 523–544
45. Ganusov, V.V., C.T. Bergstrom and R. Antia (2002), Within-host population dynamics and the evolution of microparasites in a heterogeneous host population, *Evolution*, **56**, 213–223.
46. Gao, L.Q., J. Mena-Lorca and H.W. Hethcote (1995), Four SEI endemic models with periodicity and separatices, *Math. Biosci.*, **128**, 157–184.
47. Garrett, L. (1995), *The Coming Plague*, Penguin Books, New York.

48. Gog, J.R. and J. Swinton (2002), A status-based approach to multiple strain dynamics, *J. Math. Biol.*, **44**, 169–184.

49. Grenfell, B.T. and A.P. Dobson (1995), *Ecology of Infectious Diseases in Natural Populations*, Cambridge University Press, Cambridge.

50. Hadeler, K.P. (1991), Homogeneous models in mathematical biology, *Mitt. Math. Ges. Hamburg*, **12**, 549–557.

51. Hadeler, K.P. (1992), Periodic solutions of homogeneous equations, *J. Differential Equations*, **95**, 183–202.

52. Haraguchi, Y. and A. Sasaki (2000), The evolution of parasite virulence and transmission rate in a spatially structured population, *J. Theor. Biol.*, **203**, 85–96.

53. Hatcher, M.J., D.E. Taneyhill and A.M. Dunn (1999), Population dynamics under parasitic sex ratio distortion, *Theor. Pop. Biol.*, **56**, 11–28.

54. Hethcote, H.W. (2000), The mathematics of infectious diseases, *SIAM Review* **42**, 599–653.

55. Hethcote, H.W., L.Q. Gao and J. Mena-Lorca (1996), Variations on a theme of SEI endemic models, In Martelli, M., K.L. Cooke, E. Cumberbatch, B. Tang and H.R. Thieme, *Differential Equations and Applications to Biology and Industry*, 191–207, World Scientific, Singapore.

56. Hudson, P.J., A. Rizzoli, B.T. Grenfell, J.A.P. Heesterbeek and A.P. Dobson (2002), *The Ecology of Wildlife Diseases*, Oxford University Press, Oxford.

57. Hutchinson, E. (1978), *An Introduction to Population Ecology*, Yale University Press, New Haven.

58. Iannelli, M., M. Marcheva and X.-Z. Li (2005), Strain replacement in an epidemic model with perfect vaccination, *Math. Biosci.* **195**, 23–46.

59. Lenski, R.E. and B.R. Levin (1995), Constraints on the coevolution of bacteria and virulent phage: a model, some experiments, and predictions for natural communities, *The American Naturalist* **125**, 585–602.

60. Levin, B.R. (1996), The evolution and maintenance of virulence in microparasites, *Emerging Infectious Diseases*, **2**, 93–102.

61. Levin, B.R. and R.E. Lenski (1985), Bacteria and phage: a model system for the study of the ecology and co-evolution of hosts and parasites, In D. Rollinson and R.M. Anderson *Ecology and Genetics of Host-Parasite Interactions*, 227–224, The Linnean Society of London, Academic Press.

62. Levin, B.R., F.M. Stewart and L. Chao (1977), Resource-limited growth, competition, and predation: a model and experimental studies with bacteria and bacteriophage, *Amer. Nat.*, **111**, 3–24.

63. Levin, S.A. (1970), Community equilibria and stability, and an extension of the competitive exclusion principle, *Amer. Nat.*, **117**, 413–423.

64. Levin, S.A. (1999), *Fragile Domininion: Complexity and the Commons*, Perseus Books, Reading, MA.

65. Levin, S.A. and D. Pimentel (1981), Selection of intermediate rates of increase in parasite-host systems, *Amer. Nat.*, **117**, 308–315.

66. Li, J., Z. Ma, S.P. Blythe and C. Castillo-Chavez (2003), Coexistence of pathogens in sexually-transmitted disease models, *J. Math. Biol.*, **47**, 547–568.

67. Lin, J., V. Andreasen and S.A. Levin (1999), Dynamics of influenza A drift: the linear strain model, *Math. Biosci.*, **162**, 33–51.

68. Lipsitch, M. and M.A. Nowak (1995a), The evolution of virulence in sexually transmitted HIV/AIDS, *J. Theor. Biol.*, **174**, 427–440.

69. Lipsitch, M., M.A. Nowak, D. Ebert and R.M. May (1995b), The population dynamics of vertically and horizontally transmitted parasites, *Proc. R. Soc. London*, **B 260**, 321–327.

70. Lipsitch, M., S. Siller and M.A. Nowak (1996), The evolution of virulence in pathogens with vertical and horizontal transmission, *Evolution*, **50**, 1729–1741.

71. Martcheva, M., M. Iannelli and X.-Z. Li, Competition and coexistence of strains: the impact of vaccination (preprint)

72. May, R.M. (2001), *Stability and Complexity in Model Ecosystems*, 8th printing, Princeton University Press, Princeton.

73. Metz, J.A.J., S.D. Mylius and O. Diekmann (1996), *When Does Evolution Optimise? On the Relation between Types of Density Dependence and Evolutionarily Stable Life History Parameters*, Working Paper WP-96–04, IIASA, Laxenburg.

74. Mylius, S.D. and O. Diekmann (2001), The resident strikes back: invader-induced switching of resident attractor, *J. Theor. Biol.*, **211**, 297–311.

75. Nuño, M., Z. Feng, M. Martcheva and C. Castillo-Chavez, Dynamics of two-strain influenza with isolation and partial cross-immunity, *SIAM J. Appl. Math.*, **65** (2005), 964-982.

76. O'Neill, S.L., J.H. Werren and A.A. Hoffmann (1997), *Influential Passengers: Inherited Microorganisms and Arthropod Reproduction*, Oxford Univ. Press, Oxford.

77. Pugliese, A. (2002a), Virulence evolution in macro-parasites In Castillo-Chavez, C., S. Blower, P. van den Driessche, D. Kirschner and A.-A. Yakubu, *Mathematical Approaches for Emerging and Reemerging Infectious Diseases: an Introduction*, Springer, New York, pp. 193–213

78. Pugliese, A. (2002b), On the evolutionary coexistence of parasite strains, *Math. Biosci.*, **177&178**, 355–375.

79. Rass, L. and J. Radcliffe (2003), *Spatial Deterministic Epidemics*, AMS, Providence.

80. Regoes, R.R., M.A. Nowak and S. Bonhoeffer (2000), Evolution of virulence in a heterogeneous host population, *Evolution*, **54**, 64–71.

81. Saunders, I.W. (1981), Epidemics in competition, *J. Math. Biol.* **11**, 311–318.

82. Schulthess, F.M. and S.H. Faeth (1998), Distribution, abundances, and associations of the endophytic fungal community of Arizona fescue (*Festucca arizonica*), *Mycologia*, **90**, 569–578.

83. Stearns, S.C. (1999), *Evolution in Health and Disease*, Oxford Univ. Press, Oxford, New York.

84. Tanaka, M.M. and M.W. Feldman (1999), Theoretical considerations of cross-immunity, recombination and the evolution of new parasitic strains, *J. Theor. Biol.*, **198**, 145–163.

85. Thieme, H.R. (1992), Epidemic and demographic interaction in the spread of potentially fatal diseases in growing populations, *Math. Biosci.*, **111**, 99–130.

86. Thieme, H.R. (1993), Persistence under relaxed point-dissipativity (with applications to an endemic model), *SIAM J. Math. Anal.*, **24**, 407–435.

87. Thieme, H.R. (2003), *Mathematics in Population Biology*, Princeton University Press, Princeton.

7

Directional Evolution of Virus Within a Host Under Immune Selection

Yoh Iwasa, Franziska Michor, and Martin Nowak

Summary. Viruses, such as the human immunodeficiency virus, the hepatitis B virus, the hepatitis C virus, undergo many rounds of inaccurate reproduction within an infected host. They form a heterogeneous quasispecies and change their property following selection pressures. We analyze models for the evolutionary dynamics of viral or other infectious agents within a host, and study how the invasion of a new strain affects the composition and diversity of the viral population. We previously proved, under strain specific immunity, that (Addo et al. 2003) the equilibrium abundance of uninfected cells declines during viral evolution, and that (Bittner et al. 1997) the absolute force of infection increases during viral evolution. Here we extend the results to a wider class of models describing the interaction between the virus population and the immune system. We study virus induced impairment of the immune response and certain cross-reactive stimulation of specific immune responses. For nine different mathematical models, virus evolution reduces the equilibrium abundance of uninfected cells and increases the rate at which uninfected cells are infected. Thus, in general, virus evolution tends to increase its pathogenicity. Those trends however do not hold for general cross-reactive immune responses, which introduce frequency dependent selection among viral strains. Hence an idea for combating HIV infection is to construct a virus mutant that can outcompete the existing infection without being pathogenic itself.

7.1 Introduction

Many pathogenic microbes have high mutation rates and evolve rapidly within a single infected host individual. For example, the human immunodeficiency virus (HIV) can generate mutations, and escape from immune responses and drug treatment (Hahn et al. 1986; Holmes 1992; Fenyo 1994; McMichael and Phillips 1997; Borrow et al. 1997). The continuous evolution of HIV within an infected individual over several years shifts the balance of power between the immune system and the virus in favor of the virus (Nowak et al. 1991). Virus evolution as mechanism of disease progression in HIV infection has been a common theme for the last 15 years (Nowak et al. 1990,

1995; DeBoer and Boerlijst 1994; Sasaki 1994; Nowak and May 2000). The basic theoretical idea is that a rapidly replicating HIV quasispecies establishes a permanent infection that goes through many viral generations within a short time. The immune system responds to various viral epitopes, but the virus population escapes from many such responses by generating mutants that are not recognized in particular epitopes. During the cause of infection, virus evolution proceeds toward increasing pathogenicity by reducing immune control and increasing viral abundance. There is ample experimental evidence for this mode of disease progression: (i) The HIV population in any one infected host is fairly homogeneous during primary phase but becomes heterogeneous afterwards (Bonhoeffer and Nowak 1994; Bonhoeffer et al. 1995; Wolinsky et al. 1996); (ii) the average life-cycle of HIV during the asymptomatic phase of infection is short, about 1-2 days (Ho et al. 1995; Perelson et al. 1996; Bonhoeffer et al. 1997); hence the HIV quasi-species can rapidly respond to selection pressure; (iii) HIV escapes from B-cell and T-cell mediated immune responses (Phillips et al. 1991; Wei et al. 2003; Addo et al. 2003).

In Iwasa et al. (2004), we analyze three models for the interaction between a virus population and immune responses (Perelson 1989; McLean and Nowak 1992; De Boer and Booerlijst 1994; Nowak and Bangham 1996; De Boer and Perelson 1998; Bittner et al. 1997; Perelson and Weisbuch 1997; Wodarz et al. 1999; Wahl et al. 2000; Nowak and May 2000). The models describe deterministic evolutionary dynamics in terms of uninfected cells, infected cells and strain-specific immune responses, in which there are n virus strains (or mutants) which induce n immune responses that are directed at the strains that induce them. Virus mutants can differ in all virological and immunological parameters.

In the absence of immune responses only one virus strain with the maximum fitness can survive at equilibrium. However, in the presence of immune responses, multiple strains can coexist stably. Consider a population of viral strains at a stable equilibrium. Suppose that a new strain is generated by mutation. There can be several different outcomes: the new strain may simply be added to the existing population thereby increasing the number of strains by one; the new strain may invade the existing population and other strains may become extinct; or the new strain may not be able to invade.

We ask whether there are quantities that will consistently increase (or decrease) during such viral evolution. We can prove that neither viral load nor viral diversity increases monotonically with virus evolution (although they are likely to increase in a probabilistic sense). Iwasa et al. (2004) proved that any successful invasion of a new virus strain always decreases the total abundance of uninfected cells if the immune response is specific to the strain. Further we find that any successful invasion increases the total force of infection, denoted by $\sum_{i=1}^{n} \beta_i y_i$. In the present chapter, after summarizing Iwasa et al. (2004), we mathematically examine how the invasion of a new strain affects the composition and diversity of viral population in a host for some classes of models with virus induced impairment of immune responses or cross-reactive immune stimulations. We can show that the same directional evolutionary trends as in the models without cross-immunity hold for a class of model with cross-reactive impairment or activation of immune response. Under these settings. pathogenicity always increases by evolution within a host individual.

However we can also illustrate that these unidirectional trends of virus evolution under immune selection do not hold for general cross-reactive immune responses, in which a new strain can increase the uninfected cell number.

7.2 Model of cytotoxic immunity

We start with a model in which cytotoxic immune responses reduce the lifetime of infected cells (Iwasa et al. 2004). Let x be the abundance of uninfected target cells, and y_i be the abundance of cells infected with virus strain i. Let z_i be the abundance of immune cells specific to strain i. Consider the following system of ordinary differential equations:

[Model 1] : Strain specific immunity

$$\frac{\mathrm{d}}{\mathrm{dt}}x = \lambda - dx - \sum_{i=1}^{n} \beta_i x y_i \, , \tag{1a}$$

$$\frac{\mathrm{d}}{\mathrm{dt}}y_i = (\beta_i x - a_i - p_i z_i)y_i \, , \quad i = 1, 2, 3, \ldots, n \, , \tag{1b}$$

$$\frac{\mathrm{d}}{\mathrm{dt}}z_i = c_i y_i - b_i z_i \, , \quad i = 1, 2, 3, \ldots, n \, . \tag{1c}$$

Target cells are supplied at a constant rate, λ, and die at a rate proportional to their abundance, dx. The infection rate is proportional to the abundance of uninfected and infected cells, $\beta_i x y_i$. Infected cells die at rate $a_i y_i$ because of viral cytopathicity. The immune response z_i is specific to virus strain i. The efficacy of the immune response in killing infected cells is given by p_i. Immune activity increases at a rate proportional to pathogen abundance, $c_i y_i$, and decreases at rate $b_i z_i$. We do not model the dynamics of free viral particles explicitly, but we simply assume that the number of free viral particles is proportional to the number of cells infected. This is valid because the number of free vial particles changes at a much shorter time scales than those variables in (1) (Regoes et al. 1998; Iwasa et al. 2004).

The equilibrium

The model given by (1) has a stable equilibrium. The equilibrium values of y_i and z_i can be written as functions of x, derived from (1b) and (1c). We

158   Yoh Iwasa et al.

denote these by $y_i(x)$ and $z_i(x)$ for $i = 1, 2, \ldots, n$. For given x, these values are either positive or zero.

$$y_i = \frac{b_i}{c_i p_i}[\beta_i x - a_i]_+ , \quad \text{and} \quad z_i = \frac{1}{p_i}[\beta_i x - a_i]_+ , \tag{2}$$

where $[x]_+ = x$, for $x > 0$, and $[x]_+ = 0$, for $x \leq 0$. Hence the equilibrium abundance of infected cells is a function of uninfected cell abundance x, and the total intensity of immune reaction Y. Combining $Y = \sum_{i=1}^{n} \beta_i y_i$ with (2), we have

$$Y = \sum_{i=1}^{n} \frac{\beta_i b_i}{c_i p_i}[\beta_i x - a_i]_+ , \tag{3}$$

at equilibrium. From, (2), y_i is zero for $x \leq a_i/\beta_i$, but is positive and an increasing function of x for $x > a_i/\beta_i$. The minimum level of uninfected cells required to sustain virus strain i is by a_i/β_i. On the other hand, (1a) indicates that $Y = \lambda/x - d$ holds at equilibrium.

The right hand side of (3) is a sum of increasing functions, and hence it is also an increasing function of x. Incontrast $Y = \lambda/x - d$ is a decreasing function of x. Hence there is always a single positive solution x^* at which (3) is equal to $Y = \lambda/x - d$. x^* is the equilibrium number of uninfected cells. Figure 7.1 plotted (3) and $Y = \lambda/x - d$, in which the horizontal axis is x,

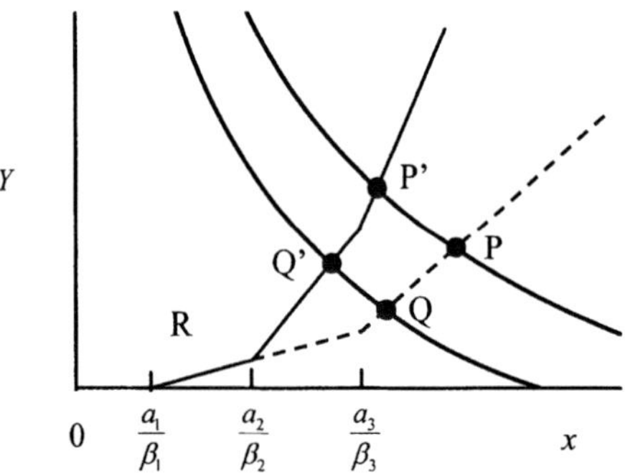

Fig. 7.1. Graphical representation of (3) and $Y = \lambda/x - d$ for a population before and after the invasion of a new strain. The model is given by (1). *Broken curve* is for the population with strain 1 and strain 3. *Solid curve* is for the population with strain 2 is added. *Three arcs connected by kink* is (3), indicating per capita risk of uninfected cells. The curves with negative slopes are $Y = \lambda/x - d$, with different value of λ. Horizontal axis is the abundance of uninfected cells x. P and Q are for the equilibrium corresponding to different λ, both including two strains. After invasion of strain 2, (3) would change to a solid curve and the equilibria would shift to P' and Q'. All three strains coexist in P'. But strain 3 is replaced by strain 2 in Q'

and the vertical axis is Y. Equation (3) is a piecewise straight line with a positive slope. $Y = \lambda/x - d$ is a curve with a negative slope. The equilibrium solution x^* is given by their cross point.

As explained in Iwasa et al. (2004), the model given by (1) has a Lyapunov function and hence the equilibrium calculated in this way is globally stable.

The possibility of invasion of a new strain into the population and its outcome is also known from a figure such as Fig. 7.1. After invasion, (3) increases by $\beta_j y_j(x)$. If, before the invasion of strain j, the population has a level of uninfected cells less than a_j/β_j, the invasion is not successful. If instead the level of uninfected cells before the invasion is greater than a_j/β_j, then strain j can increase. As an outcome of invasion, the cross–point would shift to above and toward left. The level of uninfected cells x becomes smaller than before the invasion, and Y is larger than before the invasion, and hence $Y = \sum_{i=1}^{n} \beta_i y_i$ should increase.

Figure 7.1 illustrates the situation where two strains (strain 1 and strain 3) exist in the initial population, and then strain 2 invades it ($a_1/\beta_1 < a_2/\beta_2 < a_3/\beta_3$). The broken curve in Fig. 7.1 is for the population before the invasion including strains 1 and 3 only. It consists of three arcs connected by kinks. Two curves with negative slopes are $Y = \lambda/x - d$ for different levels of λ. Both P and Q are the communities with two strains. Strain 2 with an intermediate value of a_2/β_2 is added to the population.

Consider the case in which population indicated by P is realized before the invasion of strain 2. When the strain 2 invades, the equilibrium would be shifted to P' in which all the three strains coexist because the new cross point is larger than a_i/β_i of these strains. In this case the outcome of invasion is simply the addition of a new strain 2 without extinction of the resident strains. If the population before invasion is the one indicated by Q with strains 1 and 3. The outcome of the invasion of strain 2 is the one indicated by Q' in which strains 1 and 2 coexist, but strain 3 is not maintained. This implies that the invasion of strain 2 is successful, and it drives strain 3 to extinction– the replacement of strain 3 by strain 2 occurs. The new level of uninfected cells x is too low for the strain 3 to be maintained.

From these arguments, we can see the following: (Addo et al. 2003) The invasibility of a novel strain is determined by whether or not the equilibrium abundance of uninfected cells before the invasion is greater than a_i/β_i (invasible if $x^*_{\text{before}} > a_i/\beta_i$; not invasible otherwise). (Bittner et al. 1997) As the result of a successful invasion, the location of the equilibrium would move upward and the abundance of uninfected cells downward ($x^*_{\text{after}} < x^*_{\text{before}}$). (Bonhoeffer and Nowak 1994) If x^* moves less than the threshold for some resident species $x^*_{\text{after}} < a_j/\beta_j$, they should go extinct, while those species would remain positive if $x^*_{\text{after}} > a_j/\beta_j$ is satisfied. As a result of invasion, the equilibrium intensity of immune reaction Y increases, but the number of strains maintained in the system may increase or remain unchanged or decrease. To clarify, we state this as the following proposition:

Proposition 1. *After a new strain succeeds in invasion, the equilibrium abundance of uninfected cells x always becomes less than the level before the invasion. The equilibrium total force of infection $\sum_{i=1}^{n} \beta_i y_i$ always increases after such an evolutionary change.*

A rigorous proof will be given in a later section. Before giving a formal proof, we would like to explain several different models of interaction between strains in which a similar evolutionary trend holds.

Note that the number of coexisting strains may not increase monotonically, because the invasion of a strain may cause the extinction of many existing residents. We also note that the total virus load $\sum_i y_i$ may decrease, but a properly weighted sum of viruses would increase all the time as stated in Proposition 1.

7.3 Cytotoxic immunity with proportional activation term

Next, we study another model for strain specific immunity, given by (1) in which (1c) is replaced by the following:

[Model 2]:

$$\frac{\mathrm{d}}{\mathrm{d}t} z_i = (c_i y_i - b_i) z_i , \quad i = 1, 2, 3, \ldots, n .\tag{4}$$

Here the immune response reduces the life-time of infected cells, as in model 1, but the population growth rate of immune cells specific to strain i is proportional to their current number as well as the number of infected cells: the rate of immune cell production in (4) is given by $c_i y_i z_i$ instead of $c_i y_i$ as in (1c). If viral abundance is kept constant, the immune activity shows an exponential increase in (4), but a linear increase in (1c). Again, there is a single, globally stable equilibrium (see appendix A of Iwasa et al. 2004). It is also similar to a model by Regoes et al. (1998), but parameters a_i, p_i, c_i were assumed common among strains (no suffix) in Regoes et al., but they can differ between strains in (4).

The equilibrium abundance of y_i can be expressed as a function of uninfected cell number x and the intensity of total immunity Y.

$$\text{(Case 1) for } x > \frac{a_i}{\beta_i} , \quad y_i = \frac{b_i}{c_i} , \quad z_i = \frac{\beta_i}{p_i} \left(x - \frac{a_i}{\beta_i} \right) ,\tag{5a}$$

$$\text{(Case 2) for } x = \frac{a_i}{\beta_i} , \quad 0 < y_i < \frac{b_i}{c_i} , \quad z_i = 0 ,\tag{5b}$$

$$\text{(Case 3) for } x < \frac{a_i}{\beta_i} , \quad y_i = z_i = 0 .\tag{5c}$$

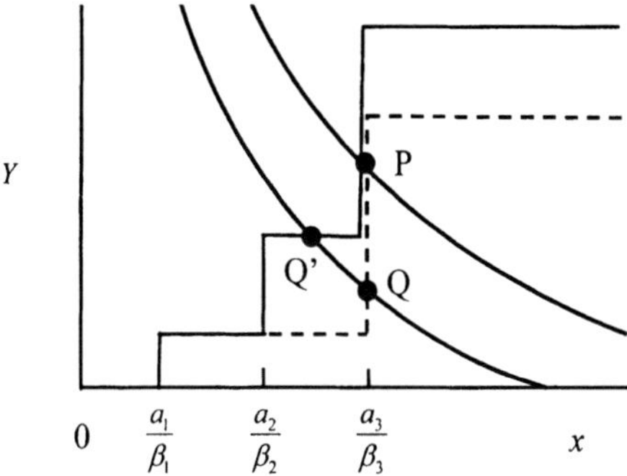

Fig. 7.2. Graphical representation of (6) and $Y = \lambda/x - d$ for a population before and after the invasion of a new strain. The model is given by (1a), (1b) and (4). Equation (6) is a step like function. *Broken curve* is for the population with strain 1 and strain 3. *Solid curve* is for the population with strain 2 is added. The curves with negative slopes are $Y = \lambda/x - d$ with different λ. Horizontal axis is the abundance of uninfected cells x. P and Q are for the equilibrium corresponding to different λ, both including two strains. After invasion of strain 2, (6) would change to a solid curve. The equilibrium P remains the same on this graph, but now includes three strains. But the uninfected cell number (horizontal axis x) does not change. In contrast Q will shift to Q', and the strain 3 is replaced by strain 2 and the equilibrium number of uninfected cell x decreases (moves toward left) after invasion

On a (x, y_i)–plane, with fixed Y, equilibrium condition (5) is represented as three straight lines with a step-like form y_i is a continuous function of x except for a single point $x = a_i/\beta_i$, at which y_i can take any value within an interval $0 < y_i < \frac{b_i}{c_i}$, which appears as a vertical line. Figure 7.2 illustrates an example. Equation (3) now becomes

$$Y = \sum_{i=1}^{n} \frac{\beta_i b_i}{c_i} H\left[x - \frac{a_i}{\beta_i}\right], \qquad (6)$$

where $H[x] = 1$, for $x \geq 0$ and $H[x] = 0$, for $x < 0$ is a Heaviside function. Equation (6) can be used except for $a_i/\beta_i (i = 1, 2, \ldots, n)$, at which one of y_i is discontinuous. When the right hand side is discontinuous ($x = a_i/\beta_i$), we can interpret (6) as indicating that Y is between the limit from below and the limit from above of the right hand side.

We assume that species differ in discontinuous points (a_i/β_i). Then there is at most one species that might cross the curve if (4) and vertical line of $x = a_i/\beta_i$, all the other species are either $x > a_i/\beta_i$ or $x < a_i/\beta_i$ at equilibrium. This requires a slight modification to Proposition 1. There can be the situation in which a new strain invades successfully and replace the resident, and yet the abundance of uninfected cells x remains exactly the same as before. Graphical representation of (6) and $Y = \lambda/x - d$ is shown in Fig. 7.2. Here equilibrium P did not change, and the equilibrium number of uninfected cells (x^*) remains the same as before. But a new strain is added without extinction of the residents. In contrast, equilibrium Q would shift to Q' after the invasion of strain 2, which causes the extinction of strain 3 and x^* becomes smaller than before. (Iwasa et al. 2004). However the equilibrium abundance of the uninfected cells should not increase after a successful invasion, it either decreases or remains unchanged. As a result, the value of $Y = \sum_{i=1}^{n} \beta_i y_i$ either increases or remains unchanged after a successful invasion, respectively. We summarize the result as follows:

Proposition 2. *If the invasion of a new strain is successful, the equilibrium abundance of uninfected cells x never decreases in the evolutionary change. It never increases. The equilibrium total force of infection $\sum_{i=1}^{n} \beta_i y_i$ either increases or remains the same as before, respectively.*

A formal proof of this proposition 2 will be given later.

7.4 Models of immune impairment

Before explaining the proof of the two propositions, we would like to explain other models that behave in a similar manner. We examine the models including the interaction between the immune reaction to different strains, such as cross-reactive immune impairment and cross–reactive immune activation, which were not covered in Iwasa et al. (2004).

[Model 3]: Cross-reactive immune impairment

Consider the model of the virus-immunity dynamics, which is composed of (1a) and (1b), but using the following, instead of (1c):

$$\frac{dz_i}{dt} = c_i y_i - b_i z_i \left(1 + u \sum_{j=1}^{n} \beta_j y_j \right). \tag{7}$$

Equation (7) indicates that the decay rate is not a constant but an increasing function of the total abundance of virus, $b_i \left(1 + u \sum_{j=1}^{n} \beta_j y_j \right)$. This assumption represents that any viral strain impairs immune activity against other viral strains. Based on a similar logic, we can prove Proposition 1 the same evolutionary trend to hold for the model given by (7), which includes cross-immunity ($u > 0$). Hence the successful invasion of a new strain always decreases the equilibrium abundance of uninfected cells, and always increases the total force of infection $\sum_{i=1}^{n} \beta_i y_i$.

[Model 4]:Same as Model 3 but with a proportional activation term

We may consider the following dynamics of immune cells,

$$\frac{dz_i}{dt} = \left(c_i y_i - b_i \left(1 + u \sum_{j=1}^{n} \beta_j y_j \right) \right) z_i. \tag{8}$$

In this model, immune cells that are specific against virus mutant i are activated at a rate, $c_i y_i z_i$, which is proportional to the product of virus abundance and immune cell abundance (Nowak and Bangham 1996). The second term within brackets of (8) implies that the mortality of immune cells increases with general activity of viral load ($u \sum_{i=1}^{n} \beta_i y_i$). This is also similar to a model by Regoes et al. (1998), but parameters a_i, p_i, c_i were assumed common among strains (no suffix) in Regoes et al., but they can differ between strains in (8).

The equilibrium abundance of y_i can be expressed as a function of uninfected cell number x and the intensity of total immunity Y.

$$\text{(Case 1) for } x > \frac{a_i}{\beta_i}, \quad y_i = \frac{b_i}{c_i}(1 + uY), \quad z_i = \frac{\beta_i}{p_i}(x - \frac{a_i}{\beta_i}) \qquad (9a)$$

$$\text{(Case 2) for } x = \frac{a_i}{\beta_i}, \quad 0 < y_i < \frac{b_i}{c_i}(1 + uY), \quad z_i = 0 \qquad (9b)$$

$$\text{(Case 3) for } x < \frac{a_i}{\beta_i}, \quad y_i = z_i = 0 \qquad (9c)$$

The graphical representation is useful. On a (x, y_i)-plane, with fixed Y, equilibrium condition (9) is represented as three straight lines with a step-like form, similar to (6). (3) now becomes

$$\frac{Y}{1 + uY} = \sum_{i=1}^{n} \frac{\beta_i b_i}{c_i} H \left[x - \frac{a_i}{\beta_i} \right]_+ . \qquad (10)$$

For this model we can prove Proposition 2. The equilibrium abundance of the uninfected cells should not increase after a successful invasion, it either decreases or remains unchanged. As a result, the value of $Y = \sum_{i=1}^{n} \beta_i y_i$ also either increases or remains unchanged after a successful invasion, respectively.

[Model 5]: Impairment of immune cell activation

Regoes et al. (1998) also consider the case in which the immune system impairment appear as a factor reducing the rate of immune activation:

$$\frac{dz_i}{dt} = \left(\frac{c_i y_i}{1 + u \sum_{j=1}^{n} \beta_j y_j} - b_i \right) z_i . \qquad (11)$$

In this model, all virus mutants contribute with different efficiency, β_j, to impairment of immune cell activation. For this model too, we can prove Proposition 2.

[Model 6]: Cross-reactive immune activation

In all the models of interaction between immune systems to different strains studied so far, the presence of a strain impairs the immune reaction of other strains. This may be plausible for HIV infection because infection of one strain would impair the general immune system.

A common way of interaction between different immune reactions is cross-immunity, in which an antigen stimulates the immune reaction of other antigens that are similar to the original one. To represent this, we consider

$$\frac{dz_i}{dt} = c_i y_i \left(1 + u \sum_{i=1}^{n} \beta_i y_i \right) - b_i z_i \, . \tag{12}$$

Here, the presence of any strain would reduce the equilibrium abundance of all the other strains. For dynamics with (1a), (1b), and (12), Proposition 1 holds. In fact, as we show later, the proof of the proposition is easier for cross-immunity models than the models with immune impairment.

[Model 7]: Cross-immunity with an alternative form

We can also consider the following form:

$$\frac{dz_i}{dt} = \left(c_i y_i \left(1 + u \sum_{i=1}^{n} \beta_i y_i \right) - b_i \right) z_i \, . \tag{13}$$

which is an alternative form of cross-immunity. For model with (1a), (1b), and (13), we can prove Proposition 2.

7.5 Proof of directional evolution

To prove the directionality of the evolutionary process, as stated in Propositions 1 and 2, we consider the following general model in which immune reaction to different strains interact. Let $Y = \sum_{i \in A} \beta_i y_i$.

$$\frac{dx}{dt} = \lambda - dx - xY \, , \tag{14a}$$

$$\frac{dy_i}{dt} = y_i f_i(x, y_i, Y, z_i) \, , \quad i = 1, 2, 3, \ldots, n \, . \tag{14b}$$

$$\frac{dz_i}{dt} = g_i(x, y_i, Y, z_i) \, , \quad i = 1, 2, 3, \ldots, n \, . \tag{14c}$$

Let A be a set of strains ($A \subset \{1, 2, 3, \ldots, n\}$). Suppose there is an equilibrium formed by a group of strains in set A. Let x^* and Y^* be the equilibrium number of uninfected cells and the total force of immunity. We further assume that, starting from any point in which all the strains in A have a positive abundance, it will converge to the equilibrium (i. e. it is globally stable).

From the dynamics given by (14b) and (14c), we can calculate y_i and z_i as a function of x and Y. In the situation for Proposition 1 to hold, such as the model given by (1), the equilibrium is a continuous function of x and Y. Here we first concentrate on such a situation (the cases in which y_i is a step function of x will be handled later). We denote the equilibrium abundance of cells infected by strain i by

$$y_i = \phi_i(x, Y) \,, \tag{15}$$

which is calculated from (14b) and (14c). In the equilibrium of the whole system (14), we have:

$$Y^* = \sum_{i \in A} \beta_i \phi_i(x^*, Y^*) \,, \tag{16}$$

from the definition of Y. From (14a), we also have

$$Y^* = \frac{\lambda}{x^*} - d \,, \tag{17}$$

at equilibrium.

Strain i has a positive abundance at equilibrium if x^* is greater than a_i/β_i, the minimum x for strain i to maintain. If the level of x^* is too high, some of the strains in set A may go extinct in the equilibrium. We have

Strain i has a positive abundance at equilibrium, if $\phi_i(x^*, Y^*) > 0$, (18a)

Strain i is absent at equilibrium, if $\phi_i(x^*, Y^*) = 0$. (18b)

In a similar manner, we can express the invasion condition in terms of ϕ. When a strain k which is not in A invades the equilibrium, whether or not it increases can be judged by the sign of $\phi_k(x^*, Y^*)$:

Strain k can invade the equilibrium, if $\phi_k(x^*, Y^*) > 0$, (19a)

Strain k fails to invade the equilibrium, if $\phi_k(x^*, Y^*) = 0$. (19b)

To discuss the outcome of a successful invasion, we assume the following two conditions:

[Condition 1] $\phi_i(x, Y)\frac{1}{Y}$ is a decreasing function of Y if $\phi_i(x, Y) > 0$.
[Condition 2] $\phi_i(x, Y)$ is a continuous and non–decreasing function of x.

All the models we have been discussed have the unique positive equilibrium satisfying (16) and (17). This can be shown, as follows: We define: $\psi(x) = (1/Y(x)) \sum_{i=1} \beta_i \phi_i(x, Y(x))$. If Y is replaced by $Y(x) = \lambda/x - d$, (16) becomes $1 = \psi(x)$. $\psi(x)$ is an increasing function of x, because $Y(x)$ is a decreasing function, and that $(1/Y) \sum_{i=1} \beta_i \phi_i(x, Y)$ is a decreasing function of Y. Note $\psi(x) = 0$ for $x \leq \min_i(a_i/\beta_i)$ because $\phi_i(x, Y) = 0$ for $x \leq a_i/\beta_i$. Also note $\lim_{Y \to +0}(1/Y) \sum_{i=1} \beta_i \phi_i(x, Y) = \infty$ for $x > \min_i(a_i/\beta_i)$. Combining these, we can conclude that there is the unique solution with $x > 0$

which satisfies both (16) and (17). Using this, we can calculate all the other variables (y_i and z_i for all i).

The global stability of this positive equilibrium is proved for models 1 and 2 in Iwasa et al. (2004), using a Lyapunov function. But for other models, we simply assume the global stability. When an invasion of mutant is successful, the positive equilibrium satisfying (16) and (17) would shift to a new positive equilibrium that is unique. This conjecture is supported by all the simulations we have done.

Under this stability assumption, we calculate the directionality of the evolution as follows (see appendix A for proof):

Theorem 1. *If Conditions 1 and 2 are satisfied, after a successful invasion of a strain, the equilibrium abundance of uninfected cells x becomes smaller than the level before the invasion. The total rate of infection, $\sum_{i \in A} \beta_i y_i x$, increases by invasion.*

Note that the increase in $\sum_{i \in A} \beta_i y_i x$ implies the increase of per capita rate of infection $Y = \sum_{i \in A} \beta_i y_i$, because x decreases by the invasion. Hence from Theorem 1, we can conclude Proposition 1.

When equilibrium y_i is a step function of x

For the model (1a), (1b) combined with immunity dynamics given by (4), (8), (11) or (13), y_i is not a continuous function of x, and hence Condition 2 is not satisfied. However y_i is expressed as (15) except for a single point $x = a_i/\beta_i$, at which y_i is not specified but takes any value between the maximum and the minimum, exemplified by (5b). We here assume that a_i/β_i differ between species. At $x = a_i/\beta_i$ ($i = 1, 2, \ldots, n$), the right hand of (16) is discontinuous. Then, we use the following inequality instead of (16):

$$\sum_{i \in A} \beta_i \phi_i(x - 0, Y) \leq Y \leq \sum_{i \in A} \beta_i \phi_i(x + 0, Y) . \tag{20}$$

We summarize these as follows:

[Condition 3] $\phi_i(x, Y)$ is a continuous and non–decreasing function of x except for a single point $x = a_i/\beta_i$, in which it is not defined. We have $\phi_i(x, Y) = 0$ for $x < a_i/\beta_i$, and $\phi_i(x, Y) > 0$ for $x > a_i > \beta_i$. At $x = a_i/\beta_i$, we have (20).

In appendix A, we can prove the following Theorem 2.

Theorem 2. *If Conditions 1 and 3 are satisfied, after a successful invasion of one or more strains, the equilibrium abundance of uninfected cells x either decreases from the level before the invasion or remains the same. The equilibrium rate of infection, $\sum_{i \in A} \beta_i y_i x$, increases or remain the same, respectively.*

In appendix B, we can show that these conditions are met for the models with (1a) and (1b), together with the immunity dynamics given by (4), (8), (11), or (13). For these models, Theorem 2 holds, and hence Proposition 2 holds, because the increase in $Y = \sum_{i \in A} \beta_i y_i$ is derived from the increase in $\sum_{i \in A} \beta_i y_i x$.

7.6 Target cells are helper T cells

HIV infects $CD4^+$ T helper cells. By depleting this target cell population, HIV impairs immune responses. In this section, we therefore assume that uninfected target cells, x, are needed for immune activation (Wodarz et al. 1999; Wodarz and Nowak 2000; Wahl et al. 2000). We consider models in which the dynamics of specific immune cells depends directly on the number of uninfected cells. Suppose immune activation requires the presence of a sufficiently many helper T cells in the tissue but the shortage of uninfected helper–T would cause the general decrease in the immune activity for all the antigens. This can be expressed as the immune activation rate dependent directly on the uninfected cell number x.

[Model 8]:

$$\frac{dz_i}{dt} = z_i(c_i y_i x - b_i), \quad i = 1, 2, \ldots, n.$$
(21)

In (21) the stimulation of immune reaction is proportional to the abundance of uninfected cells x. This was called "target cell dependence in immune activation" by Regoes et al. (1998). If a strain is abundant, it infects and reduces uninfected cell number x, which causes the decrease of the immune activation for all the other strains. Hence Regoes et al. regarded this as a way of introducing immune impairment by cross–immunity, and also called it "indirect impairment model". We can prove that, for the model with immune dynamics (21), Proposition 2 holds.

We may also think of the system in which (21) is replaced by the following:

[Model 9]:

$$\frac{dz_i}{dt} = c_i y_i x - b_i z_i, \quad i = 1, 2, \ldots, n.$$
(22)

The model, given by (1a), (1b) and (22), satisfies the condition for Theorem 1, and hence we have Proposition 1. The equilibrium abundance of uninfected cells decreases and the $Y = \sum_{i \in A} \beta_i y_i$ increases after a successful invasion of a mutant.

Bistability

In contrast, consider the case in which the target cell dependence is of impairment type, and the degree of the dependence is stronger than the one assumed by (21). For example,

$$\frac{\mathrm{d}z_i}{\mathrm{d}t} = z_i(c_i y_i x^2 - b_i) , \quad i = 1, 2, \ldots, n .$$ (23)

instead of (21). The equilibrium number of cells infected by strain i is:

$$Y = \begin{cases} \frac{b_i}{x^2 c_i} & \text{for} \quad x > a_i/\beta_i \\ 0 & \text{for} \quad x < a_i/\beta_i \end{cases} .$$ (24)

The equilibrium is determined by a solution of the following equality:

$$\lambda - dx = \frac{1}{x} \left(\frac{\beta_i b_i}{c_i} H \left[x - \frac{a_i}{\beta_i} \right] + \frac{\beta_2 b_2}{c_2} H \left[x - \frac{a_2}{\beta_2} \right] \right) ,$$ (25)

where $H[\cdot]$ is the Heaviside function. There are three equilibria – the one in the middle is unstable, and the smallest possible and the largest possible equilibria are both stable. Hence the model constituting (1a), (1b), and (23) is bistable.

7.7 General cross–immunity violates the fundamental theorem

We have been studied the evolutionary trends of virus within a host individual for a particular model of interaction between immunity to different strains. However in general cases of the cross–immunity, the decrease in the equilibrium abundance of uninfected cells no longer holds, as illustrated by two examples in Iwasa et al. (2004). One of the examples was

$$\frac{\mathrm{d}x}{\mathrm{d}t} = \lambda - dx - \sum_{i=1}^{n} \beta_i x y_i ,$$ (26a)

$$\frac{\mathrm{d}}{\mathrm{d}t} y_i = \left(\beta_i x - a_i - p_i \sum_{j=1}^{m} c_{ij} z_j \right) y_i , \quad i = 1, 2, 3, \ldots, n ,$$ (26b)

$$\frac{\mathrm{d}}{\mathrm{d}t} z_j = \sum_{j=1}^{m} y_i c_{ij} - b_j z_j , \quad i = 1, 2, 3, \ldots, m .$$ (26c)

Here i distinguishes viral strains, and j indicates epitopes. z_j is the number of immune cells specific to epitope j. The number of epitopes is m, which can be different from the number of strains n. If two strains share a common epitope, the abundance of one strain stimulates the immune reaction to the epitope and affects the other strain, which causes cross–immunity. In (26), c_{ij} is the rate of stimulation of strain i to activate the immune reaction to the jth epitope. The same matrix is used in (26b), which indicates that a strain stimulating an epitope is more likely to be suppressed by the corresponding immunity.

Iwasa et al. (2004) discussed a case of two strains and 1 epitope ($n = 2, m = 1$) with the following parameters: $\beta_1 = \beta_2 = a_1 = a_2 = 1$, $d = 0$, $c_1 = 1$, $c_2 = 5$, $p_1 = 10$, $p_2 = 1$. There is no equilibrium in which both strain 1 and strain 2 coexist. The equilibrium with strain 1 only is invaded by mutant strain 2 which replaces strain 1. The evolutionary change makes the number of uninfected cells at equilibrium 5 times greater than before. Hence the conjectured statement of monotonic decrease in uninfected cell number does not hold.

7.8 Discussion

In this paper, we studied the evolution of virus within a patient by analyzing a series of models for the dynamics of multiple strains of virus and the immune activities of the host corresponding to those strains. The immune activities to different antigens may interact with each other. We study both the case in which immune reaction to an antigen impairs the immune reaction to other antigens and the case in which the presence of an antigen stimulates the immune activity to other antigens (cross–immunity).

In most cases studied in the present paper, the directional trends of virus evolution is proved, which were shown previously for the models without cross-immunity (Iwasa et al. 2004). The equilibrium abundance of uninfected cells decreases monotonically in the viral evolution occurs within a host if controlled by immune selection. It also suggests that the total force of infection increases monotonically with the evolutionary changes of viral strain composition. The strain diversity and the mean virulence of the virus may increase statistically, but can decrease for a particular situation. In contrast the two tendencies we proved are the changes that always occur in those directions.

Regoes et al. (1998) studied by computer simulation of several different models in which the presence of a virus strain impair or suppress the immune reaction on other strains. For all the models studied by Regoes et al., we study slightly modified versions in the present paper. The modification is on the assumption of impairment function – the rate of immune activation or decay is a function of the total number of uninfected cells ($\sum_{i=1}^{n} y_i$) in Regoes et al., but the total force of infection ($\sum_{i=1}^{n} \beta_i y_i$) in the present paper. In addition, several parameters fixed by Regoes et al., can differ between strains in this paper.

Although Regoes et al. (1998) focused the case with immunity impairment, we also studied cases with cross–immunity in which a presence of one strain activate, rather than impair, the immune reaction to other strains. When cross-immunity is at work, the increase of general viral abundance should reduce the increase rate of each viral strain, and hence $y_i = \phi_i(x, Y)$

is likely to be a decreasing function of Y. Hence cross-immunity models, [Condition 1] is likely to satisfy. In contrast, for models with immune impairment has $y_i = \phi_i(x, Y)$ an increasing function of Y. If the impairment effect is very strong, Condition 1 is not satisfied, and we will not obtain the directional evolution suggested by Propositions 1 and 2. This is shown by the case with (23), which has bistability. Hence the condition for Propositions is easier to satisfy in the models with cross-immunity than in the ones with immune impairment.

Whether or not the conditions required for proposition 1 and 2 are sufficiently close to those observed in real immune systems is an important question to study in immunology. However given that there are a group of models describing the interaction between immune reaction to different strains, in which the evolution of virus population within a single patient is the monotonic increase in pathogenicity, we may be able to have a simple picture of viral evolution as a first step approximation to reality. After the infection to a host, the virus might be suppressed by the immune system to a sufficiently low level, but as the evolution progresses, the viral strains would be replaced by different strains that would cause increasingly smaller abundance of uninfected cells, and increasing higher total force of infection. Such a gloomy picture of viral evolution might be the mainstream path of the things occurring within patient of HIV.

But the mathematical result can also be used to change the direction of viral evolution. To do so, we need to produce a vaccination of a novel strain that can cause strong activation of the immune reaction, but not so much to itself. After receiving such a strain, the total force of infection by viruses would be reduced and the number of uninfected cells would recover (see Iwasa et al. 2004). Our results do not hold for general cross-reactive immunity. In this case, it is possible that viral evolution increases the equilibrium abundance of uninfected cells, reduces viral cytopathicity and reduces the force of infection. This has important implications for a completely new approach to anti–viral therapy: persistent infections in a host individual could be combated by introducing specific strains that reduce the extent of disease and/or eliminate infection (see also Bonhoeffer and Nowak 1994). An ordinary form of cross-immunity is the one in which the presence of a particular antigen enhances the immune activity to other antigens, but it may impair the immune reaction, as studied by Regoes et al. (1998).

Acknowledgement. This work was done during Y.I.'s visit to Program for Evolutionary Dynamics, Harvard University in 2003 and 2004. Program for Evolutionary Dynamics, Harvard University, is supported by Jeffrey A. Epstein.

Appendix A

Proof of Theorem 1

Let A be a group of strains with a positive abundance in the equilibrium. Let x^* and Y^* be the uninfected cell number and the total force of infection at the equilibrium. Then from (15): $\phi_i(x^*, Y^*) > 0$ for all $i \in A$. We also have

$$1 = \sum_{i \in A} \frac{1}{Y^*} \beta_i \phi_i(x^*, Y^*) \,, \tag{A.1}$$

from (16). We consider strain k, which is not in A, invades the equilibrium. From (19b), if $\phi_k(x^*, Y^*) = 0$, the invasion attempt fails. If instead

$$\phi_k(x^*, Y^*) > 0 \tag{A.2}$$

strain k increases when rare. It can invade A (see, (19a)). Then how does the abundance of uninfected cell number change after such a successful invasion? We denote $B = A \cup \{k\}$. Let x^B and Y^B be values in the new equilibrium after the invasion. Note that some of the strains in set B may have gone extinct in the new equilibrium. In the new equilibrium, (16) becomes

$$1 = \sum_{i \in A} \frac{1}{Y^B} \beta_i \phi_i(x^B, Y^B) + \frac{1}{Y^B} \beta_k \phi_k(x^B, Y^B) \,. \tag{A.3}$$

From (17), we have $Y^B = \lambda / x^B - d$. From (A.2) and (A.3), we have

$$1 > \sum_{i \in A} \frac{1}{Y^B} \beta_i \phi_i(x^B, Y^B) \,. \tag{A.4}$$

Now we can prove $x^B < x^*$, implying that the equilibrium number of uninfected cells should decrease after a successful invasion. The proof is done by assuming the opposite inequality $x^B \geq x^*$ and deriving the contraction. If $x^B \geq x^*$, we have $Y^B \leq Y^*$ from (17). From Conditions 1 and 2,

$$\left[\begin{array}{l} \text{The right hand} \\ \text{side of Eq.(A.4)} \end{array} \right] = \sum_{i \in A} \frac{1}{Y^B} \beta_i \phi_i(x^B, Y^B) \geq \sum_{i \in A} \frac{1}{Y^*} \beta_i \phi_i(x^*, Y^*) = 1 \,,$$
$$\tag{A.5}$$

where we used (A.1) for the last equality. Combing this and (A.4), we reach $1 > 1$, which is the contradiction. Hence we cannot assume $x^B \geq x^*$, and hence we conclude $x^B < x^*$.

From (17), $Yx = \lambda - dx$ holds at equilibrium. Hence the product of Y and x must increase when x decreases after the invasion of k. (End of proof of Theorem 1).

Proof of Theorem 2

Let A be a group of strains with a positive abundance in the equilibrium. Let x^* and Y^* be the uninfected cell number and the total force of infection at equilibrium. Then there are two situations:

Case 1 – For all i in A, $x^* > a_i/\beta_i$, and hence $\phi_i(x^*, Y^*) > 0$.

Case 2 – There is one strain j in A, at which $x^* = a_j/\beta_j$ holds. For all the other trains in A, $x^* > a_i/\beta_i$ and hence $\phi_i(x^*, Y^*) > 0$.

For Case 1, we can apply the same argument used to prove Theorem 1 concerning the shift in the equilibrium when an invader succeeds. Hence Theorem 1 holds, which implies Theorem 2 holds. In the following we focus on Case 2.

We denote the set of all the strains in A except for j by A'. Hence $A = A' \cup \{j\}$. We assume a similar setting as Theorem 1. Then concerning the abundance of the "boundary strain" j, we have

$$\sum_{i \in A} \frac{1}{Y^*} \beta_i \phi_i(x^*, Y^*) < 1 < \sum_{i \in A} \frac{1}{Y^*} \beta_i \phi_i(x^*, Y^*) + \frac{1}{Y^*} \beta_j \phi_j(x^* + 0, Y^*) .$$

$$(A.6)$$

Note that $\phi_j(x, Y^*)$ is discontinuous at $x = x^*$, and we need to keep $x^* + 0$ symbol indicating the limit from above. But for all the strains i in A', $\phi_i(x, Y^*)$ is continuous, which removes symbol for limit from below in (A.6).

If invader k satisfies $a_k/\beta_k > x^*$, the invasion should fail (see (19)). Invasion would be successful when $a_k/\beta_k < x^*$ and hence $\phi_k(x^*, Y^*) > 0$.

After such a successful invasion, strain j may still remain in the system at a positive abundance, or strain j may go extinct. This can be distinguished into the following two cases:

[Case 2a]   If the following inequality holds,

$$\sum_{i \in A} \frac{1}{Y^*} \beta_i \phi_i(x^*, Y^*) + \frac{1}{Y^*} \beta_k \phi_k(x^*, Y^*) < 1 , \qquad (A.7)$$

strain j still remains in the system in the new equilibrium keeping a reduced but positive abundance. Then the number of uninfected cells remains x^*, the same value as before the invasion. The outcome of the invasion is simply addition of strain k to the community. The abundances of different strains in the new equilibrium are:

$$y_i = \phi_i(x^*, Y^*) > 0 , \quad \text{for all } i \in A' , \qquad (A.8a)$$

$$y_k = \phi_k(x^*, Y^*) > 0 , \qquad (A.8b)$$

$$y_j = \frac{1}{\beta_j} \left(Y^* - \sum_{i \in A'} \beta_i \phi_i(x^*, Y^*) - \beta_k \phi_k(x^*, Y^*) \right) > 0 . \qquad (A.8c)$$

[Case 2b]   In contrast, if

$$\sum_{i \in A} \frac{1}{Y^*} \beta_i \phi_i(x^*, Y^*) + \frac{1}{Y^*} \beta_k \phi_k(x^* - 0, Y^*) > 1 , \tag{A.9}$$

strain j cannot be maintained after the invasion of strain k. In this case, we can apply a similar logic as used in deriving Theorem 1. Let $B = A' \cup \{k\}$. We assume the contrary to the inequality to prove: Suppose $x^B \geq x^*$. From (17), this leads to $Y^B \leq Y^*$. Then, we have

$$[\text{The left hand side of Eq.(A.7)}] = \sum_{i \in B} \frac{1}{Y^*} \beta_i \phi_i(x^*, Y^*)$$

$$\leq \sum_{i \in B} \frac{1}{Y^B} \beta_i \phi_i(x^B, Y^B) = 1 ,$$

which combined with (A.9) leads us to $1 > 1$, which is the contradiction. Hence we conclude $x^B < x^*$. From (17), we have $Y^B x^B > Y^* x^*$.

$$(\text{End of Proof of Theorem 2})$$

Appendix B

Here we show $\phi_i(x, Y)$ for all the models discussed in this paper. In all the models, (1a) is used for the dynamics of uninfected cells, and (1b) is adopted for the dynamics of cells infected by strain i. They differ in the dynamics of z_i immune activity specific to strain i.

Model 1 (1c):

$$\phi_i(x, Y) = \frac{b_i \beta_i}{c_i p_i} \left[x - \frac{a_i}{\beta_i} \right]_+ . \tag{B.1}$$

Model 2 (4):

$$\phi_i(x, Y) = \frac{b_i}{c_i} H \left[x - \frac{a_i}{\beta_i} \right] . \tag{B.2}$$

Model 3 (7):

$$\phi_i(x, Y) = \frac{b_i \beta_i}{c_i p_i} \left[x - \frac{a_i}{\beta_i} \right]_+ (1 + uY) . \tag{B.3}$$

Model 4 (8), and Model 5 (11):

$$\phi_i(x, Y) = \frac{b_i}{c_i} H \left[x - \frac{a_i}{\beta_i} \right] (1 + uY) . \tag{B.4}$$

Model 6 (12):

$$\phi_i(x, Y) = \frac{b_i \beta_i}{c_i p_i} \left[x - \frac{a_i}{\beta_i} \right]_+ \frac{1}{1 + uY} . \tag{B.5}$$

Model 7 (13):

$$\phi_i(x, Y) = \frac{b_i}{c_i} H \left[x - \frac{a_i}{\beta_i} \right] \frac{1}{1 + uY} . \tag{B.6}$$

Model 8 (21):

$$\phi_i(x, Y) = \frac{b_i}{c_i x} H \left[x - \frac{a_i}{\beta_i} \right] . \tag{B.7}$$

Model 9 (22):

$$\phi_i(x, Y) = \frac{b_i \beta_i}{c_i p_i x} \left[x - \frac{a_i}{\beta_i} \right]_+ . \tag{B.8}$$

For models 1, 3, 6, and 9, we can prove Theorem 1. For model 2, 4, 5, 7 and 8, together with the convention (20) at $x = a_i/\beta_i$, we can prove Theorem 2.

References

1. Addo, M.M., X.G. Yu, A.Rathod, D. Cohen, R.L. Eldridge, D. Strick, M.N. Johnston, C. Corcoran, A.G. Wurcel, C.A. Fitzpatrick, M.E.Feeney, W.R.Rodriguez, N. Basgoz, R. Draenert, D.R. Stone, C. Brander, P.J. Goulder, E.S. Rosenberg, M. Altfeld and B.D. Walker (2003), Comprehensive epitope analysis of human immunodeficiency virus type 1 (HIV-1)-specific T-cell responses directed against the entire expressed HIV-1 genome demonstrate broadly directed responses, but no correlation to viral load. *J. Virol.*, **77**.
2. Bittner, B., S. Bonhoeffer, M. A. Nowak (1997), Virus load and antigenic diversity, *Bull Math Biol.*, **59**.
3. Bonhoeffer, S. and M. A. Nowak (1994), Can live attenuated virus work as post-exposure treatment? *Immunology Today*, **16**.
4. Bonhoeffer, S., E.G. Holmes and M.A. Nowak, (1995). Causes of HIV diversity. *Nature* **376**, 125.
5. Bonhoeffer, S., R.M. May, G.M. Shaw and M.A. Nowak (1997), Virus dynamics and drug therapy. *Pro. Nat. Acad. Sci.* USA, **94**.
6. Borrow, P., H. Lewicki, X. Wei, M. S. Horwitz, N. Peffer, H. Meyers, J.A. Nelson, J.E. Gairin, B.H. Hahn, M.B. Oldstone and G. M. Shaw (1997), Antiviral pressure exerted by HIV-specific cytotoxic T lymphocytes (CTLs) during primary infection demonstrated by rapid selection of CTL escape virus, *Nature Medicine*, **3**.
7. De Boer, R. J. and A. S. Perelson (1998), Target cell limited and immune control models of HIV infection: a comparison, *J. Theor. Biol.*, **190**.

8. De Boer, R. J. and M. C. B. Boerlijst (1994), Diversity and virulence thresholds in AIDS, *Proc. Nat. Acad. Sci. USA*, **91**.

9. Fenyo, E. M. (1994), Antigenic variation of primate lentiviruses in humans and experimentally infected macaques, *Immunol. Rev.*, **140**.

10. Hahn, B. H., G. M. Shaw, M. E. Taylor, R. R. Redfield, P. D. Markham, S. Z. Salahuddin, F. Wong-Staal, R. C. Gallo, E. S. Parks and W. P. Parks (1986), Genetic variation in HTLV-III/LAV over time in patients with AIDS or at risk for AIDS, *Science* , **232**.

11. Ho, D.D., A.U. Neumann, A.S. Perelson, W. Chen, J.M. Leonard and M. Markowitz (1995), Rapid turnover of plasma virions and CO4 lymphocytes in HIV-1 infection, *Nature*, **373**.

12. Holmes, E. C., L. Q. Zhang, P. Simmonds, C. A. Ludlam and A. J. Leigh Brown (1992), Convergent and divergent sequence evolution in the surface envelope glycoprotein of HIV-1 within a single infected patient, *Proc. Natl. Acad. Sci. USA*, **89**.

13. Iwasa, Y., Michor, F., Nowak, M.A., 2004. Some basic properties of immune selection, *J. Theor. Biol.*, **229**.

14. McLean, A. R., M. A. Nowak (1992), Interactions between HIV and other pathogens, *J. Theor. Biol.*, **155**.

15. McMichael, A. J. and R. E. Phillips (1997), Escape of human immunodeficiency virus from immune control, *Ann. Rev. Immunol.*, **15**.

16. Nowak, M. A. and C. R. M. Bangham, (1996), Population dynamics of immune responses to persistent viruses, *Science*, **272**.

17. Nowak, M. A., May, R. M., Sigmund, K. (1995). Immune-responses against multiple epitopes, *J. Theor. Biol.*, **175**.

18. Nowak, M. A., R. M. Anderson, A. R. McLean, T. F. W. Wolfs, J. Goudsmit and R. M. May (1991), Antigenic diversity thresholds and the development of AIDS, *Science*, **254**.

19. Nowak, M.A., R. May, (2000), *Virus Dynamics*, Oxford.

20. Nowak, M.A., R.M. May, R.M. Anderson (1990), The evolutionary dynamics of HIV-1 quasispecies and the development of immunodeficiency disease, *AIDS*, **4**.

21. Perelson, A. S. (1989), Modeling the interaction of HIV with the immune system. In *Mathematical and Statistical Approaches to AIDS Epidemiology*, C. Castillo-Chavez, ed., Lect. Notes in Biomath. **83**, Springer–Verlag, New York.

22. Perelson, A. S. and G. Weisbuch (1997), Immunology for physicists. *Rev. Modern Phys.*, **69**.

23. Perelson, A.S., A.U. Neumann, M. Markowitz, J.M. Leonard and D.D. Ho (1996), HIV-1 dynamics in vivo: Virion clearance rate, infected cell life-span, and viral generation time. *Science*, **271**.

24. Phillips, R.E., S. Rowland-Johnes, D.F. Nixon, F.M. Gotch, J.P. Edwards, A.O. Ogunlesi, J.G. Elvin, J.A. Rothbard, D.R.M. Bangham, C.R. Rizza and A.J. McMichael (1991), Human-immunodeficiency-virus genetic-variation that can escape cytotoxic T-cell recognition, *Nature*, **354**.

25. Regoes R. R., D. Wodarz, M. A. Nowak (1998), Virus dynamics: the effect of target cell limitation and immune responses on virus evolution, *J. Theor. Biol.*, **191**.

26. Sasaki, A. (1994), Evolution of antigen drift/switching: continuously evading pathogens, *J. Theor. Biol.*, **168**.

27. Wahl, L. M., B. Bittner and M. A. Nowak (2000), Immunological transitions in response to antigenic mutation during viral infection, *Int. Immunology*, **12**.
28. Wei, X., J. M. Decker, S. Wang, H. Hui, J. C. Kappes, W. Xiaoyun, J. F. Salazar, M. G. Salazar, J. M. Kilby, M. S. Saag, N. L. Komarova, M. A. Nowak, B. H. Hahn, P. D. Kwong and G. M. Shaw (2003), Antibody neutralization and escape by HIV-1, *Nature*, **422**.
29. Wodarz, D., Lloyd, A. L., Jansen, V. A. A., Nowak, M. A. (1999). Dynamics of macrophage and T–cell infection by HIV, *J. Theor. Biol.*, **196**.
30. Wodarz, D., Nowak, M. A. (2000), CD8 memory, immunodominance and antigenic escape, *Eur. J. Immunol.*, **30**.
31. Wolinsky, S. M., Korber, B.T.M., Neumann, A.U., Daniels, M., Kunstman, K.J., Whetsell, A.J., Furtado, M.R., Cao, Y.Z., Ho, D.D., Safrit, J.T. and Koup, R.A. (1996). Adaptive evolution of human immunodeficiency virus-type 1 during the natural course of infection, *Science*, **272**.

8

Stability Analysis of a Mathematical Model of the Immune Response with Delays

Edoardo Beretta, Margherita Carletti,
Denise E. Kirschner, and Simeone Marino

8.1 Introduction

The immune system is a complex network of cells and signals that has evolved to respond to the presence of pathogens (bacteria, virus, fungi). By pathogen we mean a microbial non-self recognized as a potential threat by the host. Some pathogens preferentially survive and proliferate better inside cells (intracellular pathogens) and others are extracellular (Medzhitov et al. 2002).

The two basic types of immunity are innate and adaptive. The innate response is the first line of defense; this response targets any type of microbial non-self and is non-specific because the strategy is the same irrespective of the pathogen. Innate immunity can suffice to clear the pathogen in most cases, but sometimes it is insufficient. In fact, pathogens may possess ways to overcome the innate response and successfully colonize and infect the host.

When innate immunity fails, a completely different cascade of events ensues leading to adaptive immunity. Unlike the innate response, the adaptive response is tailored to the type of pathogen. Immune responses that clear intracellular pathogens typically involve effector cells (such as cytotoxic T cells, or CTLs) while extracellular pathogens are cleared mostly by effector molecules (e. g. antibodies) involving a different cascade of cells (such as B cells) (Janeway 2001).

There is a natural temporal kinetic that arises as part of these immune responses. The innate immune response develops first occurring on the order of minutes and hours. Adaptive immunity follows innate and occurs on the order of days or weeks. Each has an inherent delay in their development (see next section), and this timing may be crucial in determining success or failure in clearing the pathogen.

8.2 Timing of innate and adaptive immunity

Cells of innate immunity recognize highly conserved structures produced by microbial pathogens. These structures are usually shared by entire classes of pathogens (Gram-negative bacteria, for example) (Janeway et al. 2002). Once recognition occurs (Akira et al. 2001; Medzhitov et al. 2002; Takeda et al. 2003), the innate immune system is activated and ensues with very rapid kinetics (on the orders of minutes to hours).

The signals induced upon recognition by the innate immune system, in turn, stimulate and orient the adaptive immune response by controlling expression of necessary costimulatory molecules (Janeway 2002). In contrast, adaptive immunity has a tremendous capacity to recognize almost any antigenic structure (i. e., different from our gene repertoire) and because antigen receptors are generated at random (Medzhitov et al. 2000), they bind to antigens regardless of their origin (bacterial, environmental or self). Thus, the adaptive immune system responds to pathogens only after they have been recognized by the innate system (Fearon et al. 1996; Janeway 1989; Medzhitov et al. 1997). It takes at least 3 to 5 days for sufficient numbers of adaptive immune cells to be produced (expansion) (Medzhitov et al. 2000).

Another delay beyond the recognition and expansion phase occurs due to activation and differentiation phases. To complete these phases cells have to circulate and traffick from the lymphatic system through blood to the site of infection (Guermonprez et al. 2002; Zinkernagel 2003). This process takes at least few days (Jenkins et al. 2001).

It is clear timing is a key step in defining immune responses. The time frame for adaptive immunity to efficiently clear a pathogen at a site of infection is generally from 1–3 weeks (Janeway 2001; Jenkins et al. 2001; Lurie 1964). Depending on different factors, such as the type of pathogen, its proliferation rate (virus, bacteria) and tropism (intracellular or extracellular), either faster or slower responses are elicited (Antia et al. 2003; Guermonprez et al. 2002; Harty et al. 2000; Wong et al. 2003). It could be more rapid if memory cells exist (Murali-Krishna et al. 1999; Sprent et al. 2002; Surh et al. 2002; Swain et al. 1999).

The ability to mount an adaptive immune response also allows hosts to recall pathogens they have already encountered, termed a memory response. This facilitates a stronger and more efficient adaptive response whenever a second infection occurs (Sprent et al. 2002). The process of vaccination exploits this idea.

Although several examples exist in the literature of DDE modeling in biosciences and in immunology (see Murray 2002), little research in the experimental setting addresses the specific timing and functional form (kernels) of these kinetics. To begin to study these questions, we first developed a general model of the two-fold immune response, specifically to intracellular bacterial pathogens, incorporating mathematical delays for both innate and adaptive immune response.

Our baseline model tracks five variables: uninfected target cells (X_U), infected cells (X_I), bacteria (B), and phenomenological variables capturing innate (I_R) and adaptive (A_R) immunity. Uninfected target cells (1) have a natural turnover (s_U) and half-life ($\mu_{X_U} X_U$) and can become infected (mass-action term $\alpha_1 X_U B$).

$$\frac{\mathrm{d}X_U}{\mathrm{d}t} = s_U - \alpha_1 X_U B - \mu_{X_U} X_U \,, \tag{1}$$

Infected cells (2) can be cleared by the adaptive response (mass-action term $\alpha_2 X_I A_R$) or they die (half-life term $\mu_{X_I} X_I$). Here the adaptive response is represented to target intracellular bacteria.

$$\frac{\mathrm{d}X_I}{\mathrm{d}t} = \alpha_1 X_U B - \alpha_2 X_I A_R - \mu_{X_I} X_I \tag{2}$$

The bacterial population (3) has a net proliferation term, represented by a logistic function $\left(\alpha_{20} B \left(1 - \frac{B}{\sigma}\right)\right)$ and is also cleared by innate immunity (mass-action term $\alpha_3 B I_R$).

$$\frac{\mathrm{d}B}{\mathrm{d}t} = \alpha_{20} B \left(1 - \frac{B}{\sigma}\right) - \alpha_3 B I_R \,. \tag{3}$$

Both innate and adaptive responses ((4) and (5), respectively) have a source term and a half-life term.

$$\frac{\mathrm{d}I_R}{\mathrm{d}t} = s_{I_R} + \int_{-\tau_1}^{0} w_1(s) f_1(B(t+s), I_R(t+s))\, \mathrm{d}s - \mu_{I_R} I_R \tag{4}$$

$$\frac{\mathrm{d}A_R}{\mathrm{d}t} = s_{A_R} + \int_{-\tau_2}^{0} w_2(s) f_2(B(t+s), A_R(t+s))\, \mathrm{d}s - \mu_{A_R} A_R \tag{5}$$

For the innate response, the source term (s_{I_R}) includes a wide range of cells involved in the first wave of defense of the host (such as natural killer cells, polymorphonuclear cells, macrophages and dendritic cells). For the adaptive response, the source term (s_{A_R}) represents memory cells that are present, derived from a previous infection (or vaccination). A zero source implies that this is the first infection with this pathogen (i. e. no memory cells exist). Both responses are enhanced and sustained by signals that we have captured by bacterial load. The amount and type of bacteria present and the duration of infection likely determine the strength and type of immune response.

Two delays are included in the model. The delay for innate immunity, τ_1, occurs on the order of minutes to hours and τ_2 is the delay for adaptive immunity on the order of days to weeks. We assume that both responses are dependent solely on the bacterial load in the previous τ_i time units $(i = 1, 2)$ where the kernel functions $w_i(s)$, $(i = 1, 2)$ weight the past values of the bacterial load $B(s)$, i. e.:

Case 1

$$f_1(B(t+s), I_R(t+s)) = B(t+s), \quad s \in [-\tau_1, 0]$$

180   Edoardo Beretta et al.

and

$$f_2\left(B\left(t+s\right), A_R\left(t+s\right)\right) = B\left(t+s\right), \quad s \in \left[-\tau_2, 0\right].$$

In a second case, we could consider a different form of the delay. For innate immunity equation we consider the interaction (mass action product) of the bacterial load and the innate response and for adaptive immunity equation the interaction of the adaptive response with infected cells (in the previous τ_i time units, $i = 1, 2$). Therefore

Case 2

$$f_1\left(B\left(t+s\right), I_R\left(t+s\right)\right) = k_{I_R} B\left(t+s\right) I_R\left(t+s\right)$$

and

$$f_2\left(X_I(t+s), A_R(t+s)\right) = k_{A_R} X_I(t+s) A_R(t+s)$$

where k_{I_R} and k_{A_R} are scaling factors. We also consider two different types of functions for the kernels $w_i(s)(i = 1, 2)$, namely exponential or uniform.

As no experimental studies explore delays in any quantitative way, little evidence is available to inform us about the shapes of the delay kernels. However, we explore two biologically plausible cases. In the case of a uniform kernel we assume that the immune response (both innate and adaptive) is uniformly dependent on the previous τ_i time units. This implies that the bacterial load over the entire infection equally influences the response (in the 2nd delay case it implies that the interaction between the response and bacteria equally influences the response).

In the case of an exponential growth kernel, we assume that both immune responses place significant emphasis on the most recent bacterial load and that the influence of bacterial load prior to the most recent history is less significant (in the 2nd delay case it implies that only the most recent history of the interaction between the response and bacteria influences the respective response).

We hypothesis that the shorter the time delay is, the less informative (uniform) is the past history of the infection. Moreover, more recent levels of infection (for example, the number of bacteria in the host in the last few of days) will likely elicit a stronger adaptive immunity response (exponential growth). This leads to our use of a uniform kernel for innate immunity and an exponential growth kernel for adaptive immunity.

To complete the development of the mathematical model, we must estimate values for the parameters and initial conditions, as well as define units. In many cases, previously published data in the literature suggest large ranges in parameter choices: we chose an average value for our model. The values of initial conditions and parameter values are given in Table 8.1 and Table 8.2,

Table 8.1. Initial conditions (cells or bacteria per cm^3 of tissue)

Name	Value	Range
$X_U(0) = X_U^{\text{baseline}}$	$1e^4$	$1e^4 - 1e^5$
$X_I(0) = X_I^{\text{baseline}}$	0	
$B(0) = B_0$	20	
$I_R(0) = I_R^{\text{baseline}}$	$1e^3$	$1e^3 - 1e^4$
$A_R(0) = A_R^{\text{baseline}}$	$1e^2$ OR 0 (for first infections)	$1e^2 - 1e^3$

respectively. Given baseline levels and half-life terms, values of source terms s_U, s_{I_R} and s_{A_R} are determined by the following conditions:

$$s_U \equiv \mu_{X_U} X_U^{\text{baseline}}, \quad s_{I_R} \equiv \mu_{I_R} I_R^{\text{baseline}}, \quad s_{A_R} \equiv \mu_{A_R} A_R^{\text{baseline}}.$$

Thus, for example, changing I_R^{baseline} will affect only s_{I_R} and not μ_{I_R}. To properly define the integrals of equations (4)–(5) (both in delay case 1 and case 2), we need the following initial conditions:

$$\begin{cases} X_I(t) \equiv 0 & \text{for} \quad t \in [-\tau_2, 0] \\ B(t) \equiv B(0) & \text{for} \quad t \in [-\tau_2, 0] \\ I_R(t) \equiv I_R^{\text{baseline}} & \text{for} \quad t \in [-\tau_1, 0] \\ A_R(t) \equiv A_R^{\text{baseline}} & \text{for} \quad t \in [-\tau_2, 0] \end{cases} \tag{6}$$

Although our model is developed to model human infection regardless of its location, we use a volumetric measure unit (i. e., number of cells per cm^3 of tissue) to possibly compare our results with available experimental data, especially in the respiratory tract and the lung (Holt 2000; Holt et al. 2000; Marino and Kirschner 2004; Marino et al. 2004; Mercer et al. 1994; Stone et al. 1992; Wigginton et al. 2001).

In this work we have analytically analyzed only case 1 of the model leaving the analysis of case 2 for a future paper.

The model yields a boundary equilibrium, corresponding to the healthy or uninfected state, and an interior equilibrium, corresponding to an infection scenario. In Sect. 8.3 we have analyzed the main mathematical properties (positivity, boundedness and permanence of solutions) of the model, but a special emphasis is devoted to the local stability analysis (see Sect. 8.4) of the equilibria and particularly of the interior equilibrium. This special attention is due to the fact that the model equations involve distributed delays over finite intervals, and not simply fixed delays or delays over infinite intervals (for which last case the characteristic equation contains the Laplace transform of the delay kernels). Therefore the characteristic equation is dependent on the choice of the delay kernels used in the model, which in the present model, are either uniform or exponential.

Table 8.2. Parameter values

Name	Definition	Range	Units	Reference
μ_{X_U}	Half-life of X_U (like macrophages)	0.011	1/day	(Van Furth et al. 1973)
α_1	Rate of infection	$1e^{-3}$	$B(t)^{-1}$/day	Estimated
α_2	Rate of killing of X_I due to A_R	$1e^{-3}$	$A_R(t)^{-1}$/day	(Flesch and Kaufmann 1990; Lewinsohn et al. 1998; Silver et al. 1998a; Tan et al. 1997; Tsukaguchi et al 1995)
μ_{X_I}	Half-life of X_I	0.011	1/day	(Van Furth et al. 1973)
α_{20}	Growth rate of B	.5	1/day	(North and Izzo 1993; Silver et al. 1998a; Silver et al. 1998b)
σ	Max # of bacteria (threshold)	1e5	$B(t)$	Estimated
α_3	Rate of killing of B due to I_R	$1e^{-4}$	$I_R(t)^{-1}$/day	(Flesch and Kaufmann 1990)
α_4	Rate of killing of B due to A_R	$1e^{-4}$	$A_R(t)^{-1}$/day	Estimated
μ_{I_R}	Half-life of innate immunity cells	.11	1/day	(Sprent et al. 1973)
μ_{A_R}	Half-life of adaptive immunity cells	0.3333	1/day	(Sprent et al. 1973)
τ_1	Delay of innate immunity	[.1, 10]	day	
τ_2	Delay of adaptive immunity	[5, 40]	day	

Furthermore, since the interior equilibrium has components dependent on the range of the delay intervals τ_i, $i = 1, 2$, the characteristic equation will result in a polynomial transcendental equation of exponential type with polynomial coefficients that are dependent on the delay (range) τ_i. In the following of this paper the range of the delay intervals τ_i, $i = 1, 2$, will simply be called the delays τ_1 and τ_2 respectively for innate and adaptive immunity response.

For the polynomial transcendental characteristic equation mentioned above, a geometric stability switch criterion has been derived that enables study of possible stability switches as functions of delays (see Beretta and Kuang 2002). In Sect. 8.5, using the parameter values of Table 8.2 and initial conditions in (6) and Table 8.1, we describe the numerical simulations of the solutions of our model for delay values close to the stability switch values.

A discussion of the mathematical results and of their biological implications for the model is presented in Sects. 8.6 and 8.7.

8.3 Analytical results

The equations of the model are:

$$\frac{\mathrm{d}X_U(t)}{\mathrm{d}t} = s_U - \alpha_1 X_U(t)B(t) - \mu_{X_U} X_U(t)$$

$$\frac{\mathrm{d}X_I(t)}{\mathrm{d}t} = \alpha_1 X_U(t)B(t) - \alpha_2 X_I(t)A_R(t) - \mu_{X_I} X_I(t)$$

$$\frac{\mathrm{d}B(t)}{\mathrm{d}t} = \alpha_{20} B(t)\left(1 - \frac{B(t)}{\sigma}\right) - \alpha_3 B(t)I_R(t) \tag{7}$$

$$\frac{\mathrm{d}I_R(t)}{\mathrm{d}t} = s_{I_R} + \int_{-\tau_1}^{0} w_1(\theta)B(t+\theta)\,\mathrm{d}\theta - \mu_{I_R} I_R(t)$$

$$\frac{\mathrm{d}A_R(t)}{\mathrm{d}t} = s_{A_R} + \int_{-\tau_2}^{0} w_2(\theta)B(t+\theta)\,\mathrm{d}\theta - \mu_{A_R} A_R(t)$$

In the following we denote by

$$\Delta(\tau_i) = \int_{-\tau_i}^{0} w_i(\theta)\,\mathrm{d}\theta\,, \quad i = 1, 2\,. \tag{8}$$

We now discuss the main mathematical properties of system (7).

Let $h = \max\{\tau_1, \tau_2\} = \tau_2$ and define

$$x(t) := (X_U(t), X_I(t), B(t), I_R(t), A_R(t)) \in \mathbb{R}^5$$

and $X_t(\theta) = X(t+\theta)\,, \quad \theta \in [-h, 0]$ for all $t \geq 0$. Then (7) can be rewritten as

$$x'(t) = F(x_t) \tag{9}$$

with initial conditions at $t = 0$ given by

$$\Phi \in C([-h, 0], \mathbb{R}^5)$$

where $C([-h, 0], \mathbb{R}^5)$ is the Banach space of continuous functions mapping the interval $[-h, 0]$ into $\mathbb{R}^5$ equipped with the (supremum) norm

$$\| \Phi \| = \sup_{\theta \in [-h, 0]} |\Phi(\theta)|$$

where $|\cdot|$ is any norm in $\mathbb{R}^5$.

For the biological relevance, according to (6) and Table 8.1, we define non-negative initial conditions

$$\Phi(\theta) \geq 0\,, \quad \theta \in [-h, 0]$$

with

$$\Phi_i(0) > 0\,, \quad i = 1, 3, 4, 5 \quad \text{and} \quad \Phi_2(0) = X_I(0) = 0$$

to (7).

184   Edoardo Beretta et al.

Lemma 1. *Any solution $x(t) = x(\Phi, t)$ of (7) with $\Phi(\theta) \geq 0, \theta \in [-h, 0]$, $\Phi(0) > 0$ (except for $\Phi_2(0) = 0$) remains positive whenever it exists, i. e. $x(t) \in \mathbb{R}_+^5$ where*

$$\mathbb{R}_+^5 = \{x = (x_1.x_2, x_3, x_4, x_5) \in \mathbb{R}^5 | x_i > 0, i = 1, 2, 3, 4, 5\}$$

Proof. Consider the third equation in (7):

$$\frac{\mathrm{d}B}{\mathrm{d}t} = B(t) \left[\alpha_{20} \left(1 - \frac{B(t)}{\sigma} \right) - \alpha_3 I_R(t) \right]$$

with $B(0) = \Phi_3(0) > 0$. Then

$$B(t) = B(0) \exp \left\{ \int_0^t \left[\alpha_{20} \left(1 - \frac{B(s)}{\sigma} \right) - \alpha_3 I_R(s) \right] \mathrm{d}s \right\} > 0, \quad t \geq 0 \quad (10)$$

The first equation in (7) gives:

$$\frac{\mathrm{d}X_U}{\mathrm{d}t} > -X_U(t)(\alpha_1 B(t) + \mu_{X_U}), \quad X_U(0) = \Phi_1(0) > 0,$$

i. e.

$$X_U(t) > X_U(0) \exp \left\{ -\int_0^t [\alpha_1 B(s) + \mu_{X_U}] \mathrm{d}s \right\} > 0, \quad t \geq 0 \quad (11)$$

Since $X_U(t) > 0, B(t) > 0$ for $t \geq 0$, the second equation in (7) gives

$$\frac{\mathrm{d}X_I}{\mathrm{d}t} > -X_I(t) (\alpha_2 A_R(t) + \mu_{X_I}), \quad X_I(0) = \Phi_2(0) = 0,$$

i. e.

$$X_I(t) > 0 \quad t \geq 0 \tag{12}$$

Consider the last two equations in (7). Since $B(\theta) = \Phi_3(\theta) \geq 0$ in $[-h, 0]$ and $B(t) > 0$ for $t \geq 0$, we have

$$\frac{\mathrm{d}I_R(t)}{\mathrm{d}t} \geq s_{I_R} - \mu_{I_R} I_R(t), \quad I_R(0) = \Phi_4(0) > 0,$$

i. e.

$$I_R(t) \geq I_R(0) e^{-\mu_{I_R} t} + \frac{s_{I_R}}{\mu_{I_R}} (1 - e^{-\mu_{I_R} t}) > 0, \quad t \geq 0. \quad (13)$$

Similarly,

$$\frac{\mathrm{d}A_R(t)}{\mathrm{d}t} \geq s_{A_R} - \mu_{A_R} A_R(t), \quad A_R(0) = \Phi_5(0) > 0,$$

i. e.

$$A_R(t) \geq A_R(0) e^{-\mu_{A_R} t} + \frac{s_{A_R}}{\mu_{A_R}} (1 - e^{-\mu_{A_R} t}) > 0, \quad t \geq 0. \quad (14)$$

This completes the proof of positivity. $\qquad\square$

Let us consider the boundedness of solutions.

Lemma 2. *Any solution $x(t) = x(\Phi, t)$ of (7) is bounded.*

Proof. Because of positivity of solutions, the first two equations in (7) give

$$\frac{\mathrm{d}}{\mathrm{d}t}(X_U(t) + X_I(t)) < s_U - \mu(X_U(t) + X_I(t))$$

where

$$\mu = \min\{\mu_{X_U}, \mu_{X_I}\}.$$

Hence

$$\limsup_{t\to\infty}(X_U(t) + X_I(t)) \le \frac{s_U}{\mu}. \tag{15}$$

Positivity of solutions still implies

$$\frac{\mathrm{d}B(t)}{\mathrm{d}t} \le \alpha_{20}B(t)\left(1 - \frac{B(t)}{\sigma}\right)$$

and therefore

$$\limsup_{t\to\infty} B(t) \le \sigma. \tag{16}$$

Accordingly, there exists a $T_\epsilon > 0$ such that for all $t > T_\epsilon + h$ ($h = \max\{\tau_1, \tau_2\}$) and for sufficiently small $\epsilon > 0$, $B(t) < \sigma + \epsilon$. Hence, the last two equations in (7) give

$$\frac{\mathrm{d}I_R(t)}{\mathrm{d}t} < s_{I_R} + (\sigma + \epsilon)\Delta(\tau_1) - \mu_{I_R}I_R(t)$$

$$\frac{\mathrm{d}A_R(t)}{\mathrm{d}t} < s_{A_R} + (\sigma + \epsilon)\Delta(\tau_2) - \mu_{A_R}A_R(t)$$

thus implying (by letting $\epsilon \to 0$),

$$\limsup_{t\to\infty} I_R(t) \le \frac{s_{I_R} + \sigma\Delta(\tau_1)}{\mu_{I_R}} \tag{17}$$

$$\limsup_{t\to\infty} A_R(t) \le \frac{s_{A_R} + \sigma\Delta(\tau_2)}{\mu_{A_R}} \tag{18}$$

This proves boundedness. $\qquad\square$

Definition 1 (Permanence of (7)). *System (7) is permanent (or uniformly persistent) if there exist positive constants $m, M, m < M$, independent of initial conditions and such that for solutions of (7), we have:*

$$\max\left\{\limsup_{t\to\infty} X_U(t), \limsup_{t\to\infty} X_I(t), \limsup_{t\to\infty} B(t), \limsup_{t\to\infty} I_R(t),\right.$$

$$\left. \limsup_{t\to\infty} A_R(t)\right\} \le M \tag{19}$$

$$\min\left\{\liminf_{t\to\infty} X_U(t), \liminf_{t\to\infty} X_I(t), \liminf_{t\to\infty} B(t), \liminf_{t\to\infty} I_R(t),\right.$$

$$\left. \liminf_{t\to\infty} A_R(t)\right\} \ge m$$

Lemma 3. *Provided that*

$$\alpha_{20} > \alpha_3 \frac{s_{I_R} + \sigma \Delta(\tau_1)}{\mu_{I_R}}, \tag{20}$$

system (7) is permanent.

Proof. Let us consider the "lim sup" i.e. the first of (19).

From the first equation of (7) and Lemma 1 (positivity), we have:

$$\frac{\mathrm{d}X_U}{\mathrm{d}t} \leq s_U - \mu_{X_U} X_U(t),$$

which implies that

$$\limsup_{t\to\infty} X_U(t) \leq \frac{s_U}{\mu_{X_U}} := \bar{X}_U. \tag{21}$$

From (16), (21) and the second of equations (7), for sufficiently large $t > 0$ and small $\epsilon > 0$, we have

$$\frac{\mathrm{d}X_I(t)}{\mathrm{d}t} < \alpha_1 \left(\frac{s_U}{\mu_{X_U}} + \epsilon \right) (\sigma + \epsilon) - \mu_{X_I} X_I(t),$$

which gives

$$\limsup_{t\to\infty} X_I(t) \leq \frac{\alpha_1 \left(\dfrac{s_U}{\mu_{X_U}} \right) \sigma}{\mu_{X_I}} := \bar{X}_I. \tag{22}$$

Hence, from (16)–(18), (21), (22) we have

$$\left(\limsup_{t\to\infty} X_U(t), \limsup_{t\to\infty} X_I(t), \limsup_{t\to\infty} B(t), \limsup_{t\to\infty} I_R(t), \limsup_{t\to\infty} A_R(t) \right)$$
$$\leq (\bar{X}_U, \bar{X}_I, \bar{B}, \bar{I}_R, \bar{A}_R) \tag{23}$$

where

$$\bar{B} = \sigma, \quad \bar{I}_R = \frac{s_{I_R} + \sigma \Delta(\tau_1)}{\mu_{I_R}}, \quad \bar{A}_R = \frac{s_{A_R} + \sigma \Delta(\tau_2)}{\mu_{A_R}}.$$

If we choose

$$M = \max(\bar{X}_U, \bar{X}_I, \bar{B}, \bar{I}_R, \bar{A}_R),$$

then there exists $M > 0$ such that the first inequality in (19) holds true.

Consider now the "liminf", i.e. the second in (19).

(i)   Consider A_R, I_R.

From (13) and (14) we have

$$\liminf_{t\to\infty} I_R(t) \geq \frac{s_{I_R}}{\mu_{I_R}} := \underline{I}_R, \quad \liminf_{t\to\infty} A_R(t) \geq \frac{s_{A_R}}{\mu_{A_R}} := \underline{A}_R \tag{24}$$

(ii) Consider B.

From (23), for sufficiently large $t > 0$ and small $\epsilon > 0$ we have

$$\frac{\mathrm{d}B(t)}{\mathrm{d}t} \geq \alpha_{20}B(t)\left(1 - \frac{B(t)}{\sigma}\right) - (\alpha_3\bar{I}_R + \epsilon)B(t)$$

$$= \alpha_{20}B(t)\left(1 - \frac{\alpha_3\bar{I}_R + \epsilon}{\alpha_{20}} - \frac{B(t)}{\sigma}\right).$$

Hence, letting $\epsilon \to 0$

$$\liminf_{t\to\infty} B(t) \geq \left(\frac{\alpha_{20} - \alpha_3\bar{I}_R}{\alpha_{20}}\right)\sigma := \underline{B} \tag{25}$$

where $\underline{B} > 0$ provided that

$$\alpha_{20} > \alpha_3\frac{s_{I_R} + \sigma\Delta(\tau_1)}{\mu_{I_R}}(= \alpha_3\bar{I}_R).$$

(iii) Consider X_U.

For large $t > 0$, small $\epsilon > 0$ we have

$$\frac{\mathrm{d}X_U(t)}{\mathrm{d}t} \geq s_U - (\alpha_1(\bar{B} + \epsilon) + \mu_{X_U})X_U$$

from which, letting $\epsilon \to 0$

$$\liminf_{t\to\infty} X_U(t) \geq \frac{s_U}{\alpha_1\sigma + \mu_{X_U}} := \underline{X}_U \tag{26}$$

(iv) Consider X_I.

For large $t > 0$, small $\epsilon > 0$ we have

$$\frac{\mathrm{d}X_I(t)}{\mathrm{d}t} \geq \alpha_1(\underline{X}_U - \epsilon)(\underline{B} - \epsilon) - (\alpha_2(\bar{A}_R + \epsilon) + \mu_{X_I})X_I(t)$$

from which, letting $\epsilon \to 0$, we obtain

$$\liminf_{t\to\infty} X_I(t) \geq \frac{\alpha_1\underline{X}_U\underline{B}}{\alpha_2\bar{A}_R + \mu_{X_I}} := \underline{X}_I \tag{27}$$

hence, provided that (20) hods true, (24)–(27) imply that

$$\left(\liminf_{t\to\infty} X_U(t), \liminf_{t\to\infty} X_I(t), \liminf_{t\to\infty} B(t), \liminf_{t\to\infty} I_R(t),\right.$$
$$\left.\liminf_{t\to\infty} A_R(t)\right) \geq (\underline{X}_U, \underline{X}_I, \underline{B}, \underline{I}_R, \underline{A}_R), \tag{28}$$

where the constants on the right side of (28) are positive.

Thus, if we choose

$$m = \min(\underline{X}_U, \underline{X}_I, \underline{B}, \underline{I}_R, \underline{A}_R), \quad m > 0$$

considering that $\liminf x(t) \leq \limsup x(t)$, we have found two positive constants $m, M, m \leq M$ such that (19) hold true. $\square$

Remark 1. As we will see in Theorem 1, if $\Delta(\tau_1) = 0$ the permanence condition (20) becomes the existence condition of the positive equilibrium E_P. Furthermore, if $\Delta(\tau_1) = 0$, then $\underline{B} = B^*$.

Concerning the equilibria of (7), we can give the following result (we omit the computations which can be easily checked):

Theorem 1. *The system (7) gives two non-negative equilibria:*

1. for all parameter values the boundary equilibrium exists

$$E_B = \left(X_U^* = \frac{s_U}{\mu_{X_U}}, X_I^* = 0\,, B^* = 0, I_R^* = \frac{s_{I_R}}{\mu_{I_R}}, A_R^* = \frac{s_{A_R}}{\mu_{A_R}} \right) \quad (29)$$

on the boundary of the positive cone in R^5 *and*
2. for $\alpha_{20} - \alpha_3 \left(s_{I_R}/\mu_{I_R} \right) > 0$ *the positive equilibrium exists*

$$E_P = (X_U^*, X_I^*, B^*, I_R^*, A_R^*) \,.$$

with the following values for each component

$$E_P = \left(\begin{array}{ll} X_U^* = \dfrac{s_U}{\alpha_1 B^* + \mu_{X_U}}, & X_I^* = \dfrac{\alpha_1 B^* X_U^*}{\alpha_2 A_R^* + \mu_{X_I}}, \quad B^* = \dfrac{\alpha_{20} - \alpha_3 \dfrac{s_{I_R}}{\mu_{I_R}}}{\dfrac{\alpha_{20}}{\sigma} + \alpha_3 \dfrac{\Delta(\tau_1)}{\mu_{I_R}}} \\[3em] I_R^* = \dfrac{s_{I_R} + \Delta(\tau_1) B^*}{\mu_{I_R}}, & A_R^* = \dfrac{s_{A_R} + \Delta(\tau_2) B^*}{\mu_{A_R}} \end{array} \right)$$

$$(30)$$

which is interior to the positive cone in R^5.

We observe that the positive equilibrium E_P exists whenever the parameter

$$R_0 := \alpha_{20} - \alpha_3 \frac{s_{I_R}}{\mu_{I_R}} \quad (31)$$

is positive and E_P coincides with the boundary equilibrium E_B as $R_0 = 0$. When $R_0 < 0$ we have only the boundary equilibrium E_B.

8.4 Characteristic equation and local stability

System (7) linearized around any of the equilibria gives

$$\frac{\mathrm{d}x(t)}{\mathrm{d}t} = Lx(t) + \int_{-h}^{0} K(\theta)x(t+\theta)\,\mathrm{d}\theta \,. \quad (32)$$

If we define by $x(t) = \mathrm{col}\left(X_U(t), X_I(t), B(t), I_R(t), A_R(t)\right)$, then by inspection of (7) we get that $L \in \mathrm{R}^{5\times 5}$ is the matrix

$$
L = \begin{pmatrix}
-\alpha_1 B^* - \mu_{X_U} & 0 & -\alpha_1 X_U^* & 0 & 0 \\
\alpha_1 B^* & -\alpha_2 A_R^* - \mu_{X_I} & \alpha_1 X_U^* & 0 & -\alpha_2 X_I^* \\
0 & 0 & \left(\alpha_{20} - \alpha_3 I_R^* - \frac{2\alpha_{20}}{\sigma} B^*\right) & -\alpha_3 B^* & 0 \\
0 & 0 & 0 & -\mu_{I_R} & 0 \\
0 & 0 & 0 & 0 & -\mu_{A_R}
\end{pmatrix}
\tag{33}
$$

and $K\left(\theta\right) : [-h, 0] \to \mathrm{R}^{5\times 5}$ is the matrix function

$$
K = \begin{pmatrix}
0 & 0 & 0 & 0 & 0 \\
0 & 0 & 0 & 0 & 0 \\
0 & 0 & 0 & 0 & 0 \\
0 & 0 & \tilde{w}_1(\theta) & 0 & 0 \\
0 & 0 & w_2(\theta) & 0 & 0
\end{pmatrix}
\tag{34}
$$

where $\tilde{w}_1(\theta) = \begin{cases} w_1(\theta) & \text{in } [-\tau_1,\, 0] \\ 0 & \text{in } [-\tau_2,\, -\tau_1] \end{cases}$. The associated characteristic equation is

$$
\det\left(\lambda I - L - \int_{-h}^{0} K\left(\theta\right) \mathrm{e}^{\lambda\theta}\, \mathrm{d}\theta\right) = 0
\tag{35}
$$

where $I \in \mathrm{R}^{5\times 5}$ is the identity matrix and λ are the characteristic roots. If we define by

$$
F_i\left(\lambda\right) := \int_{-\tau_i}^{0} w_i(\theta)\,\mathrm{e}^{\lambda\theta}\, \mathrm{d}\theta\,, \quad i = 1, 2
\tag{36}
$$

then we get the following explicit structure for the characteristic equation:

$$
\begin{vmatrix}
\lambda + (\alpha_1 B^* + \mu_{X_U}) & 0 & \alpha_1 X_U^* & 0 & 0 \\
-\alpha_1 B^* & \lambda + (\alpha_2 A_R^* + \mu_{X_I}) & -\alpha_1 X_U^* & 0 & \alpha_2 X_I^* \\
0 & 0 & \lambda - \left(\alpha_{20} - \alpha_3 I_R^* - \frac{2\alpha_{20}}{\sigma} B^*\right) & \alpha_3 B^* & 0 \\
0 & 0 & -F_1\left(\lambda\right) & \lambda + \mu_{I_R} & 0 \\
0 & 0 & -F_2\left(\lambda\right) & 0 & \lambda + \mu_{A_R}
\end{vmatrix}
$$

$$
= 0
\tag{37}
$$

It is easy to check that (37) can be written as:

$$
\left[\lambda + (\alpha_1 B^* + \mu_{X_U})\right] \left[\lambda + (\alpha_2 A_R^* + \mu_{X_I})\right]
$$

$$
\det\begin{pmatrix}
\lambda - \left(\alpha_{20} - \alpha_3 I_R^* - \frac{2\alpha_{20}}{\sigma} B^*\right) & \alpha_3 B^* & 0 \\
-F_1\left(\lambda\right) & \lambda + \mu_{I_R} & 0 \\
-F_2\left(\lambda\right) & 0 & \lambda + \mu_{A_R}
\end{pmatrix} = 0\,,
$$

190   Edoardo Beretta et al.

i. e. we have three negative characteristic roots

$$\lambda_1 = -\left(\alpha_1 B^* + \mu_{X_U}\right) \quad \lambda_2 = -\left(\alpha_2 A_R^* + \mu_{X_I}\right) \quad \lambda_3 = -\mu_{A_R} \tag{38}$$

and the other characteristic roots are solution of:

$$\det \begin{pmatrix} \lambda - \left(\alpha_{20} - \alpha_3 I_R^* - \frac{2\alpha_{20}}{\sigma} B^*\right) & \alpha_3 B^* \\ -F_1\left(\lambda\right) & \lambda + \mu_{I_R} \end{pmatrix} = 0. \tag{39}$$

Thus the study of the characteristic equation (37) is reduced to the study of (39), the remaining characteristic roots being negative.

We remark that $F_2\left(\lambda\right)$ does not appear in (39), then the characteristic roots in (39) are independent of the second delay τ_2 of the model, i. e. the term $\int_{-\tau_2}^{0} w\left(\theta\right) B\left(t + \theta\right) d\theta$ does not play any role in the local stability of the equilibria. This implies that the first delay, that of innate immunity, is determinant in disease outcome. This likely follows because the adaptive response, A_R, does not feedback into (3). Recall this was one formulation of a delay, and in other works we consider others.

Regarding local stability of the boundary equilibrium, we can prove:

Theorem 2. *The boundary equilibrium E_B is:*

1. asymptotically stable if

$$\alpha_{20} - \alpha_3 \left(s_{I_R} / \mu_{I_R}\right) < 0 \,;$$

2. linearly neutrally stable if

$$\alpha_{20} - \alpha_3 \left(s_{I_R} / \mu_{I_R}\right) = 0 \,;$$

with one real vanishing characteristic root, while others characteristic roots are negative;

3. unstable (with one positive real root) if

$$\alpha_{20} - \alpha_3 \left(s_{I_R} / \mu_{I_R}\right) > 0 \,.$$

Proof. It follows immediately from (39) (since at the boundary equilibrium $E_B, B^* = 0$ and $I_R^* = \left(s_{I_R} / \mu_{I_R}\right)$ which gives two characteristic roots: one negative $\lambda = -\mu_{I_R}$ and the other equal to the threshold parameter R_0 for the existence of interior equilibrium E_P:

$$\lambda = \alpha_{20} - \alpha_3 \left(s_{I_R} / \mu_{I_R}\right) \,.$$

$\square$

We now study the local stability of the positive equilibrium E_P. Assume $R_0 > 0$.

At E_P, B^* satisfies

$$\alpha_{20} - \alpha_3 I_R^* - \frac{\alpha_{20}}{\sigma} B^* = 0$$

and therefore (39) reduces to

$$\det \begin{pmatrix} \lambda + \frac{\alpha_{20}}{\sigma} B^* & \alpha_3 B^* \\ -F_1'(\lambda) & \lambda + \mu_{I_R} \end{pmatrix} = 0 \, . \tag{40}$$

Therefore the local stability of E_P leads to the equation

$$\lambda^2 + \lambda \left(\mu_{I_R} + \frac{\alpha_{20}}{\sigma} B^* \right) + B^* \left(\mu_{I_R} \frac{\alpha_{20}}{\sigma} + \alpha_3 F_1(\lambda) \right) = 0 \tag{41}$$

where the information of the delay τ_1 is carried by $F_1(\lambda) := \int\limits_{-\tau_1}^{0} w_1(\theta) e^{\lambda \theta} \, d\theta$ and is therefore dependent on the choice of the delay kernel $w_1(\theta)$.

8.4.1 Uniform delay kernel

Since $F_1(\lambda)$ regards the delay in immune response it is reasonable, as stated in the introduction, to assume that the delay kernel w_1 is uniform, i. e.

$$w_1(\theta) = A \, , \quad \theta \in [-\tau_1, 0] \, . \tag{42}$$

Then

$$F_1(\lambda) = \frac{A}{\lambda} (1 - e^{-\lambda \tau_1}) \tag{43}$$

which is defined since $\lambda = 0$ is not a root of (41). In fact, if $\lambda = 0$ then $F_1(0) = \Delta(\tau_1)$ and (41) becomes

$$B^*(\tau_1) \left(\mu_{I_R} \frac{\alpha_{20}}{\sigma} + \alpha_3 \Delta(\tau_1) \right) \neq 0 \, , \quad \forall \tau_1 \geq 0 \, , \tag{44}$$

where by $B^*(\tau_1)$ we emphasize the dependence on delay τ_1, as it is evident from the equilibrium components (30).

Now remark that if $\tau_1 = 0$, then $F_1(\lambda) = 0$ and (41) becomes

$$\lambda^2 + \lambda \left(\mu_{I_R} + \frac{\alpha_{20}}{\sigma} B^*(0) \right) + \mu_{I_R} \frac{\alpha_{20}}{\sigma} B^*(0) = 0 \, , \tag{45}$$

which has two negative roots, i. e. E_P is asymptotically stable at $\tau_1 = 0$.

We have thus the general problem to find the delay values τ_1, if they exist, at which for increasing τ_1 the stability of E_P changes or, in other words, at which E_P undergoes a stability switch.

Since $\lambda = 0$ cannot be a root of (41) for any $\tau_1 \geq 0$ a stability switch for E_P can only occur at delay values τ_1 at which a pair of pure imaginary roots $\lambda = \pm i\omega(\tau_1), \omega(\tau_1) > 0$, crosses the imaginary axis.

Substituting (43) in (41) it is easy to check that (41) takes the form

$$P(\lambda, \tau_1) + Q(\lambda, \tau_1)e^{-\lambda \tau_1} = 0 \tag{46}$$

where P is a third order degree polynomial

$$P(\lambda, \tau_1) = p_3(\tau_1)\lambda^3 + p_2(\tau_1)\lambda^2 + p_1(\tau_1)\lambda + p_0(\tau_1) \tag{47}$$

with delay dependent coefficients

$$\begin{cases} p_3(\tau_1) = 1 \\ p_2(\tau_1) = \mu_{I_R} + \dfrac{\alpha_{20} B^*(\tau_1)}{\sigma} \\ p_1(\tau_1) = \dfrac{\mu_{I_R} \alpha_{20} B^*(\tau_1)}{\sigma} \\ p_0(\tau_1) = \alpha_3 A B^*(\tau_1) \end{cases} \tag{48}$$

and Q is a zeroth order polynomial

$$Q(\lambda, \tau_1) = q_0(\tau_1) = -\alpha_3 A B^*(\tau_1). \tag{49}$$

The occurrence of stability switches for equations with delay dependent coefficients of the type (46) has been recently studied by Beretta and Kuang (2002) who have proposed a geometric stability switch criterion. We summarize it below.

We consider the class of characteristic equations of the form

$$P(\lambda, \tau) + Q(\lambda, \tau)e^{-\lambda \tau} = 0, \quad \tau \in \mathbb{R}_{+0} \tag{50}$$

where P, Q are two polynomials in λ

$$P(\lambda, \tau) = \sum_{k=0}^{n} p_k(\tau)\lambda^k ; \quad Q(\lambda, \tau) = \sum_{k=0}^{m} q_k(\tau)\lambda^k, \quad n, m \in \mathbb{N}_0, \ n > m \tag{51}$$

with coefficients $p_k(\cdot), q_k(\cdot) \colon \mathbb{R}_{+0} \to \mathbb{R}$ which are continuous and differentiable functions of τ.

We assume that

(H1) $P(0, \tau) + Q(0, \tau) = p_0(\tau) + q_0(\tau) \neq 0, \forall \tau \in \mathbb{R}_{+0}$ i.e. $\lambda = 0$ is not a root of (50);

(H2)   at $\tau = 0$ all roots of (50) have negative real parts;

(H3)   if $\lambda = i\omega, \omega \in \mathbb{R}$, then

$$P(i\omega, \tau) + Q(i\omega, \tau) \neq 0, \quad \forall \tau \in \mathbb{R}_{+0}.$$

We now turn to the problem of finding the roots $\lambda = \pm i\omega, \omega \in \mathbb{R}_+$ of (50).

A necessary condition is that

$$F(\omega, \tau) = 0 \tag{52}$$

where

$$F(\omega, \tau) := |P(i\omega, \tau)|^2 - |Q(i\omega, \tau)|^2 . \tag{53}$$

Let $\omega = \omega(\tau), \tau \in I \subset \mathbb{R}_{+0}$ be a solution of (52). We assume that $\omega = \omega(\tau)$ is a continuous and differentiable function of $\tau \in I$.

For each solution $\omega = \omega(\tau), \tau \in I$, of (50) we find the angle $\theta = \theta(\tau), \tau \in I$ satisfying

$$\begin{cases} \sin\theta(\tau) = \dfrac{-P_R(i\omega, \tau)Q_I(i\omega, \tau) + P_I(i\omega, \tau)Q_R(i\omega, \tau)}{|Q(i\omega, \tau)|^2} \\[2mm] \cos\theta(\tau) = -\dfrac{P_R(i\omega, \tau)Q_R(i\omega, \tau) + P_I(i\omega, \tau)Q_I(i\omega, \tau)}{|Q(i\omega, \tau)|^2} \end{cases} \tag{54}$$

where, thanks to $(H3)$ we can prove that $\theta(\tau) \in (0, 2\pi), \ \tau \in I$ and $\theta(\tau)$ is a continuous and differentiable function of $\tau \in I$.

By the functions $\omega = \omega(\tau), \theta = \theta(\tau), \tau \in I$, we define the functions $S_n: I \to \mathbb{R}$ according to

$$S_n(\tau) := \tau - \frac{\theta(\tau) + n2\pi}{\omega(\tau)}, \quad n \in \mathbb{N}_0, \tag{55}$$

which are continuous and differentiable for $\tau \in I$.

Finally, by any mathematical software such as Maple or Matlab, we draw the curves S_n versus $\tau \in I$ looking for their zeros

$$\tau^* \in I: S_n(\tau^*) = 0. \tag{56}$$

In fact, we can prove the following:

Theorem 3. *All the roots $\lambda = \pm i\omega(\tau), \omega(\tau) > 0$ of (50) occur at the delay values τ^* if and only if τ^* is a zero of one of the functions in the sequence $S_n, \ n \in \mathbb{N}_0$.*

At each $\tau^* \in I$ a pair of roots of (50) $\lambda = \pm i\omega(\tau^*)$ is crossing the imaginary axis according to the sign of

$$\mathrm{sign}\left\{\left.\frac{\mathrm{d}\mathrm{Re}\lambda}{\mathrm{d}\tau}\right|_{\lambda=\pm i\omega(\tau^*)}\right\} = \mathrm{sign}\left\{F_\omega^{'}(\omega(\tau^*), \tau^*)\right\}\mathrm{sign}\left\{\left.\frac{\mathrm{d}S_n(\tau)}{\mathrm{d}\tau}\right|_{\tau=\tau^*}\right\}. \tag{57}$$

A stability switch occurs at $\tau = \tau^* \in I$ if the total multiplicity on the right side of the imaginary axis changes from 0 to 2 or from 2 to 0 when τ increases through τ^*.

Theorem 4. *If τ^* is the lowest positive zero of the function $S_0(\tau)$ and the transversality condition*

$$sign\left\{\left.\frac{\mathrm{d}\mathrm{Re}\lambda}{\mathrm{d}\tau}\right|_{\lambda=\pm i\omega(\tau^*)}\right\} = 1$$

holds, (50) has

(a) *all roots with negative real parts if $\tau \in [0, \tau^*)$;*
(b) *a pair of conjugate pure imaginary roots $\pm i\omega(\tau^*), \omega(\tau^*) > 0$, crossing the imaginary axis, and all the other roots with negative real part if $\tau = \tau^*$;*
(c) *two roots with strictly positive real part if $\tau > \tau^*$;*
(d) *because of (b), all the roots λ $(\neq \pm i\omega(\tau^*))$ satisfy the condition $\lambda \neq im\omega(\tau^*)$, where m is any integer, if $\tau = \tau^*$.*

Hence, at $\tau = \tau^*$ a Hopf bifurcation occurs (see Hale and Verduyn Lunel, chap. 11, (Hale and Verduyn Lunel 1993))

Using the parameter values of Table 8.2 in the Introduction we can prove the following:

Theorem 5. *If the uniform delay kernel w_1 in (46)–(49) is such that $A = \dfrac{1}{\tau_1}$, then in the biological range $[0.1, 10]$ there is one stability switch at the delay value*

$$\tau_{1_0}^+ = 5.6491$$

toward instability, which is also a Hopf bifurcation value.

Proof. We describe the algorithm presented in the previous pages applied to the characteristic equation (46) whose structure is defined in (47), (48) and (49).

1st Step. From (47)–(49) we have

$$P(i\omega, \tau_1) = \left(p_0(\tau_1) - \omega^2 p_2(\tau_1)\right) + i\left(\omega p_1(\tau_1) - \omega^3\right) \tag{58}$$

with

$$P_R(i\omega, \tau_1) = p_0(\tau_1) - \omega^2 p_2(\tau_1), \quad P_I(i\omega, \tau_1) = \omega p_1(\tau_1) - \omega^3 \tag{59}$$
$$Q(i\omega, \tau_1) = q_0(\tau_1) \tag{60}$$

with

$$Q_R(i\omega, \tau_1) = q_0(\tau_1), \quad Q_I(i\omega, \tau_1) = 0. \tag{61}$$

Then

$$F(\omega, \tau_1) := |P(i\omega, \tau_1)|^2 - |Q(i\omega, \tau_1)|^2$$

yields

$$F(\omega, \tau_1) = \omega^2[\omega^4 + a_2(\tau_1)\omega^2 + a_1(\tau_1)] = 0 \tag{62}$$

where

$$\begin{cases} a_2(\tau_1) = \mu_{I_R}^2 + \left(\dfrac{\alpha_{20} B^*(\tau_1)}{\sigma}\right)^2 > 0 \\[2mm] a_1(\tau_1) = p_1^2(\tau_1) - 2p_0(\tau_1) p_2(\tau_1) \\[2mm] \qquad\quad = \left[\dfrac{\mu_{I_R} \alpha_{20} B^*(\tau_1)}{\sigma}\right]^2 - 2\alpha_3 A B^*(\tau_1)\left(\mu_{I_R} + \dfrac{\alpha_{20} B^*(\tau_1)}{\sigma}\right) \end{cases} \tag{63}$$

It is easy to check that $a_1(\tau_1) < 0$ in $[0, \tau_1^*)$ where

$$\tau_1^* := \frac{2\alpha_3 \left(\mu_{I_R} + \dfrac{\alpha_{20}B^*}{\sigma} \right)}{\left(\dfrac{\mu_{I_R}\alpha_{20}}{\sigma} \right)^2 B^*} = 1.6950 \times 10^5$$

where $B^* = 4.376 \times 10^2$ (we further note that $a_1(\tau_1^*) = 0$ and $a_1(\tau_1) > 0$ in $(\tau_1^*, +\infty)$, i.e. $F(\omega, \tau_1) > 0$ in $(\tau_1^*, +\infty)$ and no stability switch can occur in such a delay range). Since in the biological range $[0.1, 10]$ it is $a_1(\tau_1) < 0$, the only positive root of (62) in the biological range is

$$\begin{cases} \omega_+ (\tau_1) = \left[\dfrac{1}{2} \left(-a_2 (\tau_1) + \sqrt{a_2^2 (\tau_1) - 4a_1 (\tau_1)} \right) \right]^{1/2} \\ \tau_1 \in (0, \tau_1^*) = \mathrm{I} \end{cases} \tag{64}$$

(such that $\omega_+(\tau_1^*) = 0$). Furthermore, it's easy to check that

$$F_\omega' \left(\omega_+(\tau_1), \tau_1 \right) > 0 . \tag{65}$$

2nd Step. According to (59), (61) we can define the angle $\theta_+(\tau_1)$ as solution of

$$\begin{cases} \sin\theta_+ (\tau_1) = -\dfrac{\omega_+ p_1 (\tau_1) - \omega_+^3}{|q_0 (\tau_1)|} \\ \cos\theta_+ (\tau_1) = \dfrac{p_0 (\tau_1) - \omega_+^2 p_2 (\tau_1)}{|q_0 (\tau_1)|} \end{cases} . \tag{66}$$

3rd Step. We define the functions $S_n^+ : \mathrm{I} \to \mathbb{R}$, where $I = [0.1, 10]$, by

$$S_n^+ (\tau_1) := \tau_1 - \frac{\theta_+ (\tau_1) + n2\pi}{\omega_+ (\tau_1)} , \quad \tau_1 \in I = (0, \tau_1^*) , \quad n \in \mathbb{N}_0 . \tag{67}$$

According to Theorem 2 if $\tau_{1_i}^+$ is a zero of $S_n^+(\tau_1)$ for some $n \in \mathbb{N}_0$, then at $\tau_1 = \tau_{1_i}^+$ there is a pair of pure imaginary roots $\lambda = \pm i\omega_+(\tau_{1_i}^+)$, $\omega_+(\tau_{1_i}^+) > 0$, crossing the imaginary axis according to

$$\mathrm{sign} \left\{ \frac{\mathrm{dRe}\lambda}{\mathrm{d}\tau_1} \bigg|_{\lambda=\pm i\omega(\tau_{1_i}^+)} \right\} = \mathrm{sign} \left\{ F_\omega' \left(\omega_+(\tau_{1_i}^+), \tau_{1_i}^+ \right) \right\} \mathrm{sign} \left\{ \frac{\mathrm{d}S_n^+ (\tau_1)}{\mathrm{d}\tau_1} \bigg|_{\tau_1=\tau_{1_i}^+} \right\}$$

$$= \mathrm{sign} \left\{ \frac{\mathrm{d}S_n^+ (\tau_1)}{\mathrm{d}\tau_1} \bigg|_{\tau_1=\tau_{1_i}^+} \right\} . \tag{68}$$

In Fig. 8.1 are depicted the graphs of functions $S_n^+(\tau_1)$ versus τ_1 in the biological range $[0.1, 10]$. Only S_0^+ has a zero at $\tau_{1_0}^+ = 5.6491$ which, according to Theorem 2, is a stability switch delay value toward instability. Furthermore, thanks to Theorem 4 we can say that at $\tau_{1_0}^+ = 5.6491$ we have a Hopf bifurcation delay value. $\qquad\square$

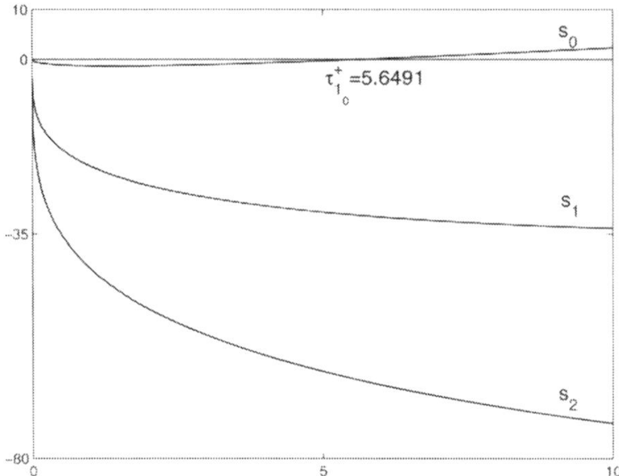

Fig. 8.1. Graphs of functions $S_n^+(\tau_1)$ versus τ_1 in the biological range $[0.1, 10]$. Only S_0^+ has one zero at $\tau_{1_0}^+ = 5.6491$ which is a stability switch delay value toward instability and a Hopf bifurcation delay value

It may be mathematically interesting to consider the functions $S_n^+(\tau_1)$ versus τ_1 even outside of the biological range. As shown in Fig. 8.2 in the range $[0.1, 350]$ for τ_1 the functions S_0^+, S_1^+, S_2^+ present zeros respectively at the delay values $\tau_{1_0}^+ = 5.6491$, $\tau_{1_1}^+ = 91.2267$ and $\tau_{1_2}^+ = 260.4919$. However, by (68) we can see that only $\tau_{1_0}^+$ is a stability switch delay value since between $\tau_{1_0}^+, \tau_{1_1}^+$ the total multiplicity of characteristic roots with positive real part, say

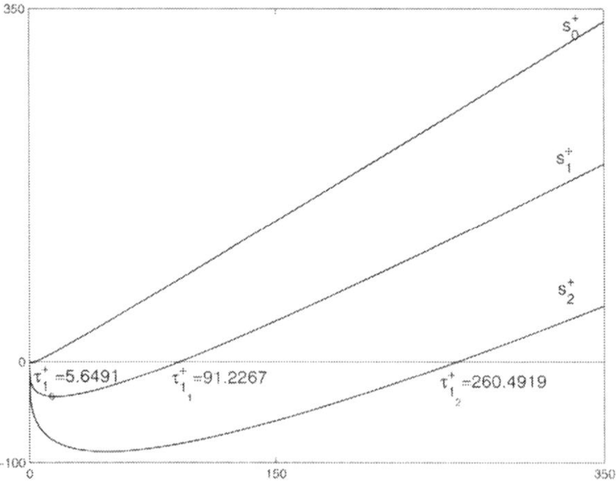

Fig. 8.2. In the range $[0.1, 350]$ are depicted the graphs of the functions S_n^+ versus τ_1. Besides S_0^+ even S_1^+ and S_2^+ have zeros respectively at $\tau_{1_1}^+ = 91.2267$ and $\tau_{1_2}^+ = 260.4919$ but the stability switch occurs at the zero of S_0^+: $\tau_{1_0}^+ = 5.6491$

ρ, is $\rho = 2$ and becomes $\rho = 4$ between $\tau_{1_1}^+$ and $\tau_{1_2}^+$ and finally becomes $\rho = 6$ beyond $\tau_{1_2}^+$. Thus, E_P becomes unstable after $\tau_{1_0}^+$ and it remains unstable on the whole range.

However, we may further observe that, since $\tau_1 \to \tau_1^*$ from left implies $\omega_+(\tau_1) \to 0$, by (66) we see that $\theta_+(\tau_1) \to 2\pi$ and by (67) we have $S_n^+(\tau_1) \to -\infty$ as $\tau_1 \to \tau_1^*$ for each $n \in \mathbb{N}_0$.

Since S_n^+ are continuous and continuously differentiable functions of τ_1, for each S_n^+ which has a zero with positive slope, there exists another one with negative slope. In conclusion, there are only two stability switches which are two external zeros of S_0^+ and which are the external zeros of all the zeros in the sequence S_n^+. The first stability switch from asymptotic stability to instability is at $\tau_{1_0}^+$ and the second stability switch is at the last zero of S_0^+, say at $\tau_{2_0}^+$ from instability to asymptotic stability.

In the interval $(\tau_{1_0}^+, \tau_{2_0}^+)$ the positive equilibrium is unstable. For $\tau_1 > \tau_{2_0}^+$ the positive equilibrium regains its asymptotic stability which is kept for all $\tau_1 \in (\tau_{2_0}^+, +\infty)$.

8.4.2 Exponential delay kernel

If in (41) we assume an exponential delay kernel

$$w_1(\theta) = A \ e^{k\theta} \ , \quad \theta \in [-\tau_1, 0] \ , \quad A, k \in \mathbb{R}_+ \tag{69}$$

then

$$F_1(\lambda) = \frac{A}{\lambda + k} \left[1 - e^{-(\lambda + k)\tau_1} \right] . \tag{70}$$

If $\lambda = -k$ is not a solution of (41) (if $\lambda = -k$ is a solution the E_P is asymptotically stable) substitution of (70) in (41) leads to (46) where the delay-dependent coefficients (48) are now given by

$$\begin{cases} p_3(\tau_1) = 1 \\[2mm] p_2(\tau_1) = k + \mu_{I_R} + \dfrac{\alpha_{20} B^*(\tau_1)}{\sigma} \\[2mm] p_1(\tau_1) = k \left(\mu_{I_R} + \dfrac{\alpha_{20} B^*(\tau_1)}{\sigma} \right) + \dfrac{\mu_{I_R} \alpha_{20} B^*(\tau_1)}{\sigma} \\[2mm] p_0(\tau_1) = +B^*(\tau_1) \left(\dfrac{k \mu_{I_R} \alpha_{20}}{\sigma} + \alpha_3 A \right) \end{cases} \tag{71}$$

and (49) by

$$q_0(\tau_1) = -\alpha_3 B^*(\tau_1) A e^{-k\tau_1} . \tag{72}$$

Even in this case if $\lambda = 0$ we have $F_1(0) = \Delta(\tau_1)$ and (44) shows that $\lambda = 0$ is not a characteristic root for any $\tau_1 \geq 0$. Furthermore, at $\tau_1 = 0$ (45) shows that E_P is asymptotically stable. Hence, again we can ask if increasing τ_1 in the biological range $[0.1, 10]$ there is a delay value at which a stability switch toward instability occurs.

We can follow the same procedure shown for the case of uniform delay kernel, taking into account that the coefficients of (46) are now given by (71) and (72). We omit the detailed computations of steps 1–3. According to Theorems 2 and 3 we can then prove:

Theorem 6. *In the exponential delay kernel (69) we choose* $A = k = \dfrac{\log 2}{\tau_1}$, *then in the biological range* $[0.1, 10]$ *there is one stability switch at the delay value*

$$\tau_{1_0}^{+} = 6.69310$$

toward instability, which is also a Hopf bifurcation delay value.

8.5 Numerical simulations

We simulated the system by numerically solving the differential equations using suitable numerical methods. Our aim was to confirm that the Hopf bifurcations in Theorems 4 and 5 give rise, for increasing delay τ_1, to solutions which show sustained oscillations. We used two different procedures to study the solutions of system (7) with initial conditions (6) and Table 8.1.

Considering the general case for delay equations (7) i.e. of exponential delay kernels for innate and adaptive immune responses,

$$w_i(\theta) = A_i e^{K_i \theta}, \quad \theta \in [-\tau_1, 0], \quad i = 1, 2$$
$$A_i, K_i \in \mathbb{R}_+$$

in system (7) we define the new variables

$$u_I(t) := \int_{-\tau_1}^{0} w_1(\theta) B(t + \theta) \, \mathrm{d}\theta$$
$$u_A(t) := \int_{-\tau_2}^{0} w_2(\theta) B(t + \theta) \, \mathrm{d}\theta \tag{73}$$

By the transformation $s = t + \theta$ (73) give

$$u_I(t) := \int_{t-\tau_1}^{t} w_1(s - t) B(s) \, \mathrm{d}s$$
$$u_A(t) := \int_{t-\tau_2}^{t} w_2(s - t) B(s) \, \mathrm{d}s \tag{74}$$

which is straightforward checking that they satisfy the equations

$$\frac{\mathrm{d}u_I}{\mathrm{d}t} = A_1 B(t) - A_1 e^{-K_1 \tau_1} B(t - \tau_1) - K_1 u_I(t)$$
$$\frac{\mathrm{d}u_A}{\mathrm{d}t} = A_2 B(t) - A_2 e^{-K_2 \tau_2} B(t - \tau_2) - K_2 u_A(t)$$

hence, system (7) is transformed into

$$\frac{\mathrm{d}X_U(t)}{\mathrm{d}t} = s_U - \alpha_1 X_U(t)B(t) - \mu_{X_U}X_U(t)$$

$$\frac{\mathrm{d}X_I(t)}{\mathrm{d}t} = \alpha_1 X_U(t)B(t) - \alpha_2 X_I(t)A_R(t) - \mu_{X_I}X_I(t)$$

$$\frac{\mathrm{d}B(t)}{\mathrm{d}t} = \alpha_{20}B(t)\left(1 - \frac{B(t)}{\sigma}\right) - \alpha_3 B(t)I_R(t)$$

$$\frac{\mathrm{d}I_R(t)}{\mathrm{d}t} = s_{I_R} + u_I(t) - \mu_{I_R}I_R(t) \tag{75}$$

$$\frac{\mathrm{d}A_R(t)}{\mathrm{d}t} = s_{A_R} + u_A(t) - \mu_{A_R}A_R(t)$$

$$\frac{\mathrm{d}u_I(t)}{\mathrm{d}t} = A_1 B(t) - A_1\,\mathrm{e}^{-K_1\tau_1}B(t-\tau_1) - K_1 u_I(t)$$

$$\frac{\mathrm{d}u_A(t)}{\mathrm{d}t} = A_2 B(t) - A_2\,\mathrm{e}^{-K_2\tau_2}B(t-\tau_2) - K_2 u_A(t)$$

with initial conditions given by (6) and Table 8.1, and particularly B takes i.c. on the interval $[-\tau_2, 0]$, i.e.

$$B(s) = \Phi_3(s)\,, \quad s \in [-\tau_2, 0]\,, \tag{76}$$

thus defining at $t = 0$ the initial conditions for u_I and u_A in (75) by

$$u_I(0) = \int_{-\tau_1}^{0} w_1(s)\Phi_3(s)\,\mathrm{d}s$$

$$u_A(0) = \int_{-\tau_2}^{0} w_2(s)\Phi_3(s)\,\mathrm{d}s\,. \tag{77}$$

Hence, system (7) is transformed into an equivalent system of delay differential equations (75) with fixed delay τ_1, τ_2, where "equivalent" means that such a new system has the same characteristic equation and same equilibria (regarding the original variables (X_U, X_I, B, I_R, A_R)) as the original system (7) (we leave to the reader to check it). Note that the case of uniform delay kernel for innate response I_R is simply obtained by setting $K_1 = 0$ in (75).

Such a new system (75) may be solved by any delay differential equations solver. We used the Matlab *dde23* by Shampine and Thompson. The second way is by directly approximating the solution of the distributed delay system through the trapezoidal rule for the equations and the (composite) trapezoidal quadrature formula for the integrals. The overall order of accuracy of the method is 2 (Baker and Ford 1988).

We have performed simulations for both cases of uniform (see Figs. 8.3, 8.4) and exponential (see Figs. 8.5, 8.6) delay kernels for the innate response and for τ_1 values below the Hopf threshold $\tau_{1_0}^{+}$ and for $\tau_1 > \tau_{1_0}^{+}$, showing in this last case that sustained oscillations occurred.

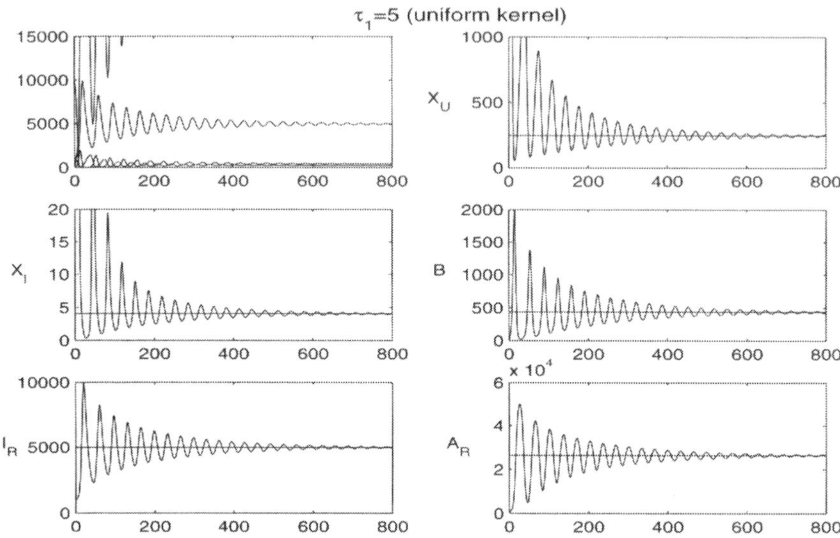

Fig. 8.3. Simulations of solutions of system (7) in case of uniform delay kernel for the innate response with $A_1 = 1$. The delay τ_1 is chosen below the threshold $\tau_{1_0}^+$ of Theorem 4. The top left figure shows the behaviour of all the variables together. The straight lines represent the equilibrium components. The delay kernel for the adaptive response is uniform with $A_2 = 1$ and τ_2 is kept fixed at the value 20

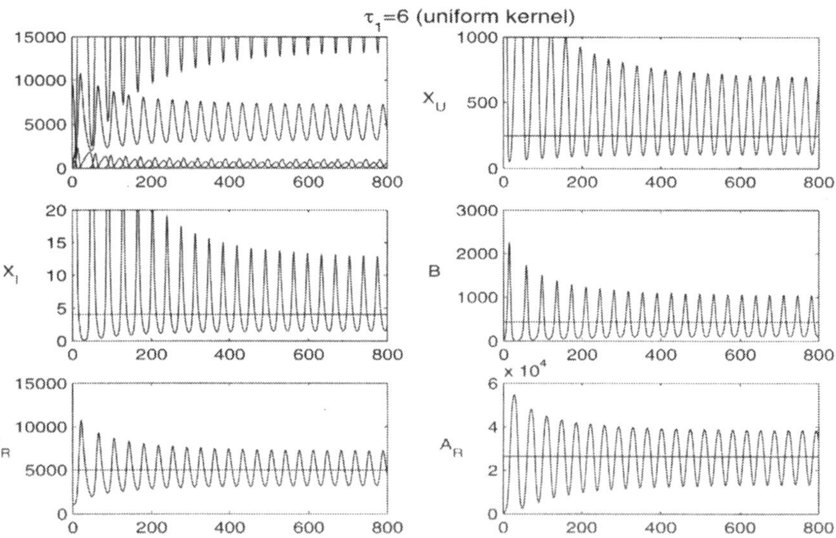

Fig. 8.4. Simulations of solutions of system (7) in case of uniform delay kernel for the innate response with $A_1 = 1$. The delay τ_1 is chosen above the threshold $\tau_{1_0}^+$ of Theorem 4. The delay kernel for the adaptive response is uniform with $A_2 = 1$ and τ_2 is kept fixed at the value 20

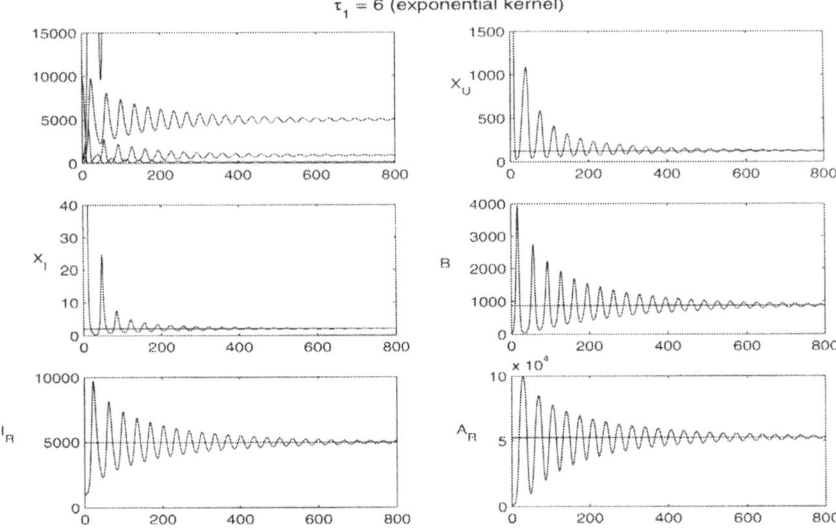

Fig. 8.5. Simulations of solutions of system (7) in case of exponential delay kernel for the innate response with $A_1 = K_1 = \log 2/\tau_1$. The delay τ_1 is chosen below the threshold $\tau_{1_0}^+$ of Theorem 5. The delay kernel for the adaptive response is uniform with $A_2 = 1$ and τ_2 is kept fixed at the value 20

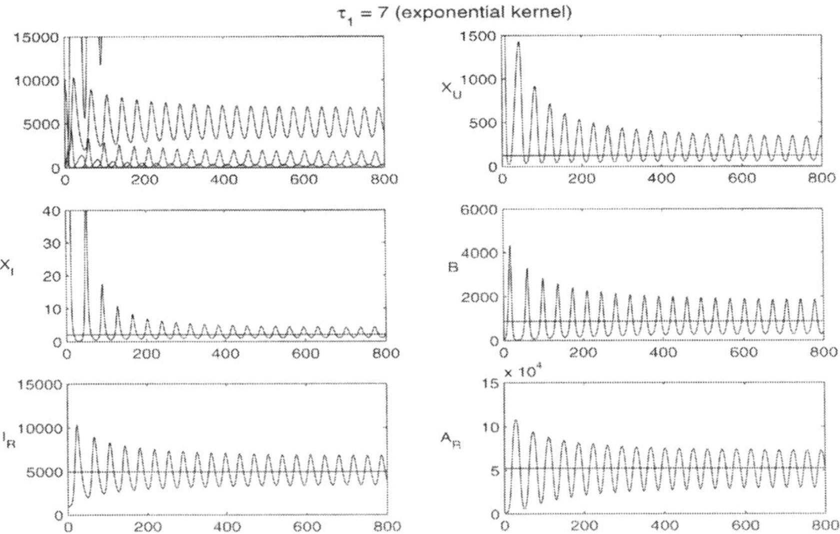

Fig. 8.6. Simulations of solutions of system (7) in case of exponential delay kernel for the innate response with $A_1 = K_1 = \log 2/\tau_1$. The delay τ_1 is chosen above the threshold $\tau_{1_0}^+$ of Theorem 5. The delay kernel for the adaptive response is uniform with $A_2 = 1$ and τ_2 is kept fixed at the value 20

8.6 Discussion

We develop a mathematical model to address timing of the immune system when challenged by intracellular bacterial infection. A baseline model accounts for different killing capabilities of the immune system and incorporates two delays representing the two types of immune responses, namely innate and adaptive immunity for which two different cases of delay interactions are proposed. We have discussed only case 1, remarking however that case 2 can be studied similarly.

The baseline model, case 1, admits a boundary equilibrium E_B or uninfected steady state and only when the threshold parameter R_0 in (31) becomes positive one positive equilibrium E_P bifurcates from E_B (transcritical bifurcation) corresponding to the infected steady state. The local stability of E_B is independent of the delays in the innate (τ_1) and adaptive (τ_2) immune responses. E_B is asymptotically stable whenever the positive equilibrium E_P is not feasible and unstable if E_P exists. The positive equilibrium E_P has components dependent on either the delay τ_1 or on the delay τ_2 although its local stability is independent of τ_2, as it is evident by (40). The study of the characteristic equation leads to (41) where the term $F_1(\lambda)$ takes information of the delay kernel $w_1(\theta)$, $\theta \in [-\tau_1, 0]$ in the innate immune response. This is a crucial point in modelling the immune system since there is little information regarding these delays. Assuming a uniform or exponential delay kernel, (41) takes the form of the polynomial exponential transcendental equation (46) with delay dependent coefficients given by (47)–(49) for uniform delay kernel or by (71)–(72) for an exponential delay kernel. Of course, the change of the numerical value of any parameter in the delay kernel may lead to different outcomes of the stability analysis. We note that at $\tau_1 = 0$ the positive equilibrium E_P is asymptotically stable whereas, increasing τ_1, E_P has a Hopf bifurcation toward sustained oscillations (see Sect. 8.5) at $\tau_{1_0}^+ = 5.6491$ in the case of uniform delay kernel (with $A = \frac{1}{\tau_1}$) or at $\tau_{1_0}^+ = 6.69310$ in the case of exponential delay kernel (with $A = K = \frac{\log 2}{\tau_1}$).

We could attempt to derive global stability results for E_P by Lyapunov functional method, but the presence of a Hopf bifurcation with respect to τ_1 should lead to severe bounds on τ_1. We do not show such computations here but the global stability result for E_P requires values of τ_1, τ_2 close to zero, i.e. useless in understanding the behaviour of the model for large delays. Though we do not discuss the baseline model case 2, it is interesting to note that this model presents many of the properties of the model case 1. In both cases there are two non-negative equilibria E_B and E_P, the second arising when the same threshold parameter R_0 in (31) is positive and again, for the stability of E_B, Theorem 2 holds true. The main difference is in the stability analysis for E_P which is now dependent on both delays τ_1 and τ_2.

8.7 Biological discussion

Our baseline model suggests a key role for innate immunity in establishing a protective response and describes how different delay times and shapes affect the pattern of bacterial growth and its impact on the host. Our study indicates how a delayed innate response (τ_1 larger than 5 days) results in oscillatory behavior, suggesting how trade offs for initial conditions of both the host (for example the baseline level of innate immunity cells, I_R^{baseline}, or the host capability of containing the early stages of infection) and the pathogen (its proliferation rate, α_{20}) determine the final infection outcome. E_B or uninfected steady state represents a successful immune response of the host: bacteria are cleared and the system returns to equilibrium. The model suggests how this scenario is stable, is readily achieved and it is independent from delays either in innate or adaptive responses. Clearance in this case seems more a structural property of the host (initial number of cells and their efficacy in killing) and the pathogenicity of the bacteria (virulence factors). E_P or the infected steady state represents successful colonization of the host by bacteria. Here the innate immunity "memory" plays an important role: the shorter τ_1, the easier infection can be stabilized (τ_1 smaller than the stability switch $\tau_{1_0}^+$). In fact, a small value for τ_1 (on the order of hours) is more biologically consistent and plausible than τ_1 on the order of days. Damped oscillation still lead to an infection scenario: some level of intracellular bacteria always persists (Figs. 8.1 and 8.3). Considering τ_1 larger than the stability switch (see Figs. 8.2 and 8.4), the average of the oscillations is equivalent to the level of bacteria in the oscillations. Although, on average, the two outcomes are similar, a biological difference can be drawn in terms of latent versus chronic infection scenarios. Latent infection represents a damped oscillation where a "peaceful" coexistence between the host and the intracellular bacteria is established. On the other hand, a chronic infection scenario is suggested by a sustained oscillation, where an "unstable" and potentially dangerous coexistence between the host and the pathogen could be driven out of control more easily by either host factor or environmental pressure. As an example consider tuberculosis infection in humans. The adaptive immune response to Mycobacterium tuberculosis infection is the formation of a multicellular immune structure called a granuloma. Recent hypotheses suggest how granuloma are a dynamic entities (Capuano et al. 2003) that contain the spread of the infection to other parts of the body. A continuous trade-off between host immune cells and bacteria numbers exists within the granuloma: waves of infection and bursting of chronically infected cells (releasing intracellular bacteria) are contained by waves of effector cells taking up bacteria and stabilizing infection. The exponential kernel induces larger oscillations in bacterial levels, suggesting how a uniform kernel is more beneficial for the innate response, and also more biologically plausible.

References

1. Akira, S., K. Takeda and T. Kaisho (2001), Toll-like receptors: critical proteins linking innate and acquired immunity, *Nat. Immunol.*, **2**, 675.
2. Antia, R., C. T. Bergstrom, S. S. Pilyugin, S. M. Kaech and R. Ahmed (2003), Models of CD8+ responses: 1. What is the antigen-independent proliferation program, *J. Theor. Biol.*, **221**, 585.
3. Baker, C. T. H. and N. Ford (1988), Convergence of linear multistep methods for a class of delay-integro-differential equations. In:*International Series of Numerical Mathematics* 86, Birkhäuser Verlag Basel.
4. Beretta, E. and Y. Kuang (2002), Geometric stability switch criteria in delay differential systems with delay dependent parameters, *SIAM J. Math. Anal.*, **33**, 1144.
5. Capuano, S. V., D. A. Croix, S. Pawar, A. Zinovik, A. Myers, P. L. Lin, S. Bissel, C. Fuhrman, E. Klein and J. L. Flynn (2003), Experimental Mycobacterium tuberculosis infection of cynomolgus macaques closely resembles the various manifestations of human M. tuberculosis infection.*Infect. Immun.*, **71**, 5831.
6. Fearon, D.T. and R. M. Locksley (1996), The instructive role of innate immunity in the acquired immune response *Science*, **272,** 50.
7. Flesch, I. E., S. H. Kaufmann (1990), Activation of tuberculostatic macrophage functions by gamma interferon, interleukin-4, and tumor necrosis factor, *Infect. Immun.*, **58**, 2675.
8. Guermonprez, P., J. Valladeau, L. Zitvogel, C. Thery, and S. Amigorena (2002), Antigen presentation and T cell stimulation by dendritic cells, *Annu. Rev. Immunol..*, **20**, 621.
9. Hale, J. K. and S. M. Verduyn Lunel (1993), *Introduction to functional differential equations*, Springer Verlag, New York.
10. Harty, J. T., A. R. Tvinnereim and D. W. White (2000), CD8+ T cell effector mechanisms in resistance to infection *Annu. Rev. Immunol.*, **18**, 275.
11. Holt, P. G. and M. A. Schon-Hegrad (1987), Localization of T cells, macrophages and dendritic cells in rat respiratory tract tissue: implications for immune function studies, *Immunology*, **62**, 349.
12. Holt, P.G. (2000), Antigen presentation in the lung, *Am. J. Respir. Crit. Care Med.*, **162**, 151.
13. Janeway, Jr. C. A. and R. Medzhitov (2002), Innate immune recognition, *Annu. Rev. Immunol.*, **20,** 197.
14. Janeway, Jr. C. A. (1989), Approaching the asymptote? Evolution and revolution in immunology, *Cold Spring Harb. Symp. Quant. Biol.*, **54 Pt 1,** 1.
15. Janeway, Jr. C. A. (2001), *Immunobiology 5 : the immune system in health and disease*, Garland Pub., New York.
16. Janeway, Jr. C. A. (2002), A trip through my life with an immunological theme *Annu. Rev. Immunol.*, **20,** 1.
17. Jenkins, M. K., A. Khoruts, E. Ingulli, D. L. Mueller, S. J. McSorley, R. L. Reinhardt, A. Itano and K.A. Pape (2001), In vivo activation of antigen-specific CD4 T cells, *Annu. Rev. Immunol.*, **19**, 23.
18. Kuang, Y. (1993), *Delay differential equations : with applications in population dynamics*, Academic Press, Cambridge, Mass.
19. Lewinsohn, D. M., T. T. Bement, J. Xu, D. H. Lynch, K. H. Grabstein, S. G. Reed and Alderson, M. R. (1998), Human purified protein derivative-specific CD4+ T cells use both CD95- dependent and CD95-independent cytolytic mechanisms, *J. Immunol.*, **160**, 2374.

20. Lurie, M. B. (1964), *Resistance to tuberculosis : experimental studies in native and acquired defensive mechanisms*, Harvard Univ. Press, Cambridge, Mass.

21. Marino, S. and D. E. Kirschner (2004), The human immune response to Mycobacterium tuberculosis in lung and lymph node, *J. Theor. Biol.* 227 **4**, 463.

22. Marino, S., S. Pawar, C. L. Fuller, T. A. Reinhart, J. L. Flynn and D. E. Kirschner, (2004), Dendritic cell trafficking and antigen presentation in the human immune response to Mycobacterium tuberculosis, *J. Immunol.* 173 **1**, 494.

23. Medzhitov R. and C. A. Janeway, Jr. (1997), Innate immunity: the virtues of a nonclonal system of recognition, *Cell*, **91**, 295.

24. Medzhitov, R. and C. A. Janeway, Jr. (2000), Innate immunity, *N. Engl. J. Med.*, **343**, 338.

25. Medzhitov, R. and C. A. Janeway, Jr. (2002), Decoding the patterns of self and nonself by the innate immune system. *Science*, **296**, 298.

26. Mercer R. R., M. L. Russell, V. L. Roggli and J. D. Crapo (1994), Cell number and distribution in human and rat airways, *Am. J. Respir. Cell. Mol. Biol.*, **10**, 613.

27. Murali-Krishna, K., L. L. Lau, S. Sambhara, F. Lemonnier, J. Altman and R. Ahmed (1999), Persistence of memory CD8 T cells in MHC class I-deficient mice, *Science*, **286**, 1377.

28. Murray, J. D. (2002), *Mathematical biology*, 3rd edn., Springer, New York.

29. North R. J. and A. A. Izzo (1993), Mycobacterial virulence. Virulent strains of Mycobacteria tuberculosis have faster in vivo doubling times and are better equipped to resist growth-inhibiting functions of macrophages in the presence and absence of specific immunity, *J. Exp. Med.*, **177**, 1723.

30. Shampine, L. F. and S. Thompson, Solving DDEs with Matlab, manuscript, URL: `http://www.radford.edu/ thompson/webddes`

31. Silver, R. F., Q. Li, J. J. Ellner (1998b), Expression of virulence of Mycobacterium tuberculosis within human monocytes: virulence correlates with intracellular growth and induction of tumor necrosis factor alpha but not with evasion of lymphocyte- dependent monocyte effector functions, *Infect. Immun.*, **66**, 1190.

32. Silver, R. F., Q. Li, W. H. Boom and J. J. Ellner (1998a), Lymphocyte-dependent inhibition of growth of virulent Mycobacterium tuberculosis H37Rv within human monocytes: requirement for CD4+ T cells in purified protein derivative-positive, but not in purified protein derivative-negative subjects, *J. Immunol.*, **160**, 2408.

33. Sprent, J. and A. Basten (1973), Circulating T and B lymphocytes of the mouse. II. Lifespan, *Cell. Immunol.*, **7**, 40.

34. Sprent, J., C. D. Surh (2002), T cell memory *Annu. Rev. Immunol.*, **20**, 551.

35. Stone, K. C., R. R. Mercer, P. Gehr, B. Stockstill and J. D. Crapo (1992), Allometric relationships of cell numbers and size in the mammalian lung, *Am. J. Respir. Cell. Mol. Biol.*, **6**, 235.

36. Surh, C. D. and J. Sprent (2002), Regulation of naive and memory T-cell homeostasis, *Microbes Infect.*, **4**, 51.

37. Swain,S. L., H. Hu and G. Huston (1999), Class II-independent generation of CD4 memory T cells from effectors *Science*, **286**, 1381.

38. Takeda, K., T. Kaisho and S. Akira (2003), Toll-like receptors, *Annu. Rev. Immunol.*, **21**, 335.

39. Tan, J. S., D. H. Canaday, W. H. Boom, K. N. Balaji, S. K. Schwander and E. A. Rich (1997), Human alveolar T lymphocyte responses to Mycobacterium tuberculosis antigens: role for CD4+ and CD8+ cytotoxic T cells and relative resistance of alveolar macrophages to lysis, *J. Immunol.*, **159**, 290.
40. Tsukaguchi, K., K. N. Balaji and W. H. Boom (1995), CD4+ alpha beta T cell and gamma delta T cell responses to Mycobacterium tuberculosis. Similarities and differences in Ag recognition, cytotoxic effector function, and cytokine production, *J. Immunol.*, **154**, 1786.
41. Van Furth, R., M. C. Diesselhoff-den Dulk and H. Mattie (1973), Quantitative study on the production and kinetics of mononuclear phagocytes during an acute inflammatory reaction, *J. Exp. Med.*, **138**, 1314.
42. Wigginton, J. E. and D. E. Kirschner (2001), A model to predict cell-mediated immune regulatory mechanisms during human infection with Mycobacterium tuberculosis, *J. Immunol.*, **166**, 1951.
43. Wong, P. and E. G. Pamer (2003), CD8 T cell responses to infectious pathogens, *Annu. Rev. Immunol.*, **21**, 29.
44. Zinkernagel, R. M. (2003), On natural and artificial vaccinations, *Annu. Rev. Immunol.*, **21**, 515

9

Modeling Cancer Treatment Using Competition: A Survey *

H.I. Freedman

Summary. Several models are proposed to simulate the treatment of cancer by various techniques including chemotherapy, immunotherapy and radiotherapy. The interactions between cancer and normal cells are viewed as competitions for resources. Using ordinary differential equations, we model these treatments as constant and periodic.

9.1 Introduction

In North America, cancer is the second largest cause of human mortality, and as such, is of great concern to the population at large. Despite the billions of dollars poured into research to date, a "cure for cancer" is still out of reach, although significant progress has been made in many types of cancers. Such progress has led to greater understanding of the cancers and their effects and in improvements in treatments leading to a better quality of life and in some cases to a cure.

Mathematics has contributed in a small way to the understanding of cancer by analysis and simulation of cancer models in a hope of discovering new insights. This is well evidenced by the publication of a special issue of the journal, *Discrete and Continuous Dynamical Systems Series B* (Horn and Webb 2004), titled "Mathematical Models in Cancer", which contains twenty-one papers concerned with modelling various types and aspects of cancer. It is interesting to note, however, that in all these works (and others) there is hardly any modelling or mention of treatment.

It is the purpose of this chapter to briefly survey how treatment may be included in cancer modelling. However, we restrict ourselves to models which treat the interactions between cancer and normal cells as a competition for bodily resources (nutrients, oxygen, space, etc.).

* Research partially supported by the Natural Sciences and Engineering Research Council of Canada, Grant No. NSERC OGP 4823.

The organization of the chapter is as follows. In Sect. 9.2 we consider our model with no treatment and state the conditions for cancer to always win. This is followed by modelling treatment by radiation using control theory in Sect. 9.3. Section 9.4 deals with chemotherapy treatment and Sect 9.5 with immunotherapy treatment. In Sect. 9.6 we look at the case where cancer metastasizes (spreads). Finally a short discussion will be in Sect. 9.7.

9.2 The no treatment case

We model the interaction between normal and cancer cells as a competition for bodily resources. Let $x_1(t)$ be the concentration of normal cells and $x_2(t)$ be the concentration of cancer cells at a given site. Then in the absence of treatment, our model takes the form

$$\dot{x}_1(t) = \alpha_1 x_1(t)\left(1 - \frac{x_1(t)}{K_1}\right) - \beta_1 x_1(t)x_2(t)\,, \qquad x_1(0) \geq 0$$

$$\dot{x}_2(t) = \alpha_2 x_2(t)\left(1 - \frac{x_2(t)}{K_2}\right) - \beta_2 x_1(t)x_2(t)\,, \qquad x_2(0) \geq 0\,, \tag{1}$$

where $\cdot = \frac{\mathrm{d}}{\mathrm{d}t}, \alpha_i$ is the proliferation coefficient, β_i is the competition coefficient and K_i is the carrying capacity for the ith cell population, $i = 1, 2$.

For this model, the following boundary (with respect to the positive quadrant) equilibria always exist, $E_0(0,0), E_1(K_1,0)$ and $E_2(0,K_2)$. It is well known (see Freedman and Waltman 1984) that for the general dynamics of solutions initiating in the nonnegative quadrant at nonequilibrium values, there are four possible outcomes, (i) x_1 always wins, (ii) x_2 always wins, (iii) there is an interior equilibrium $\widehat{E}(\widehat{x}_1, \widehat{x}_2)$, where $\widehat{x}_1 > 0, \widehat{x}_2 > 0$, and $\widehat{E}$ is asymptotically stable (and hence globally stable for strictly positive solutions), (iv) $\widehat{E}$ exists and is a saddle point, i.e. E_1 and E_2 are both locally stable, and whether x_1 or x_2 wins depends on the initial conditions.

According to our cancer assumption that cancer always wins, we require that only case (ii) occurs. Criteria for this to happen are given in Freedman and Waltman (1984), and are

$$\alpha_1 < K_2\beta_1\,, \qquad \alpha_2 > K_1\beta_2\,. \tag{2}$$

Throughout the rest of this chapter, we assume that (2) holds.

We will modify system (1) in this paper to simulate various treatments.

9.3 Treatment by radiation

The material in this section is taken (with permission) from the Masters Thesis of Belostotski (2004). In general system (1) may be modified so as to include a harvesting of cells due to radiation. The general form of the new system is then given by

$$\dot{x}_1 = \alpha_1 x_1\left(1 - \frac{x_1}{K_1}\right) - \beta_1 x_1 x_2 - \eta_1(t, x_1, x_2), \qquad x_1(0) \geq 0$$
$$\dot{x}_2 = \alpha_2 x_2\left(1 - \frac{x_2}{K_2}\right) - \beta_2 x_1 x_2 - \eta_2(t, x_1, x_2), \qquad x_2(0) \geq 0,$$

(3)

where η_i, $i = 1, 2$, is the effect of radiation on the cell populations.

In the first instance we suppose that the radiation is ideal, i.e. it targets only cancer cells. This may be effected by setting $\eta_1(t, x_1, x_2) = 0$. In the second instance we can look at the case of a minor spillover to normal cells, by writing $\eta_1(t, x_1, x_2) = \varepsilon \overline{\eta}_1(t, x_1, x_2)$, and use perturbation theory. Then at the third stage of analysis, one can consider fully system (3).

In this paper, we only consider the case where $\eta_1(t, x_1, x_2) = 0$. For the perturbation case, see Belostotski (2004). Four types of control are feasible:

$$\text{(i) } \eta_2 = \gamma = \text{const.} ; \quad \text{(ii) } \eta_2 = \gamma x_2 ; \quad \text{(iii) } \eta_2 = \gamma \frac{x_2}{x_1} ;$$

$$\text{(iv) } \eta_2 = \begin{cases} \gamma & \text{for} \quad nkT \leq t < (nk+1)T \\ 0 & \text{for} \quad (nk+1)T \leq t < (nk+2)T, \quad n \subset N. \end{cases}$$

Here we will analyze in some detail case (i). The other cases may be found in Belostotski (2004).

9.3.1 Existence of equilibria

In case (i), system (3) becomes

$$\dot{x}_1 = \alpha_1 x_1\left(1 - \frac{x_1}{K_1}\right) - \beta 1 x_1 x_2$$
$$\dot{x}_2 = \alpha_2 x_2\left(1 - \frac{x_2}{K_2}\right) - \beta_2 x_1 x_2 - \gamma .$$

(4)

Let

$$a = \alpha_1 \alpha_2 - \beta_1 \beta_2 K_1 K_2 .$$

In the absence of radiation, i.e. $\gamma = 0$, system (4) generates the following isoclines:

$$\Gamma_1 : x_1 = K_1 - \frac{\beta_1 K_1}{\alpha_1} x_2$$
$$\Gamma_2 : x_1 = \frac{\alpha_2}{\beta_2} - \frac{\alpha_2}{\beta_2 K_2} x_2 .$$

(5)

The sign of a describes the nature of the interaction between healthy and cancer cells. Consider the slopes of Γ_1 and Γ_2 in (5). If

$$\text{(i)} \quad -\frac{\alpha_2}{\beta_2 K_2} > -\frac{\beta_1 K_1}{\alpha_1} \implies a < 0\,,$$

$$\text{(ii)} \quad -\frac{\alpha_2}{\beta_2 K_2} = -\frac{\beta_1 K_1}{\alpha_1} \implies a = 0\,, \tag{6}$$

$$\text{(iii)} \quad -\frac{\alpha_2}{\beta_2 K_2} < -\frac{\beta_1 K_1}{\alpha_1} \implies a > 0\,.$$

When $a \neq 0$, the isoclines (5) do not intersect since we restrict our analysis to the case when cancer wins the competition (conditions (2)). When radiation is introduced, the equations of isoclines (5) will change to:

$$\begin{aligned}
\Gamma_1 &: x_1 = K_1 - \frac{\beta_1 K_1}{\alpha_1} x_2 \\
\Gamma_3 &: x_1 = \frac{\alpha_2}{\beta_2} - \frac{\alpha_2}{\beta_2 K_2} x_2 - \frac{\gamma}{\beta_2 x_2}\,.
\end{aligned} \tag{7}$$

Notice that on Γ_3 as $x_2 \to 0^+$, then x_1 approaches $-\infty$. In addition, on Γ_3, $\frac{dx_1}{dx_2} = -\frac{\alpha_2}{\beta_2 K_2} + \frac{\gamma}{\beta_2 x_2^2}$ and $\frac{d^2 x_1}{dx_2^2} = -\frac{2\gamma}{\beta_2 x_2^3}$. Thus Γ_3 will have the shape as depicted in Figs. 9.1 and 9.2 with the vertex (maximum value of x_1) at:

$$(x_1, x_2) = \left(\frac{\alpha_2}{\beta_2} - \frac{2}{\beta_2}\sqrt{\frac{\alpha_2 \gamma}{K_2}}\,, \sqrt{\frac{K_2 \gamma}{\alpha_2}} \right).$$

In the positive x_1, x_2 plane these isoclines may intersect twice, once, or zero times as in Figs. 9.1 and 9.2. The number of intersections depends on the size of γ and the dynamics of the cancer-healthy tissue interaction represented by a.

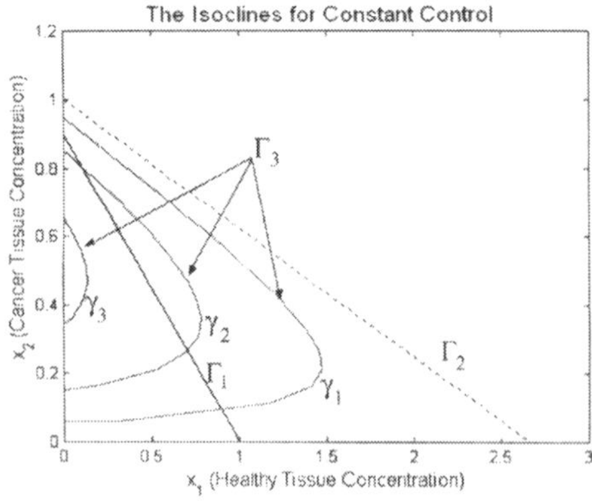

Fig. 9.1. Isoclines of (6): $a < 0$. Changes in shape of Γ_3 for different values of γ : $\gamma_1 < \gamma_2 < \gamma_3$

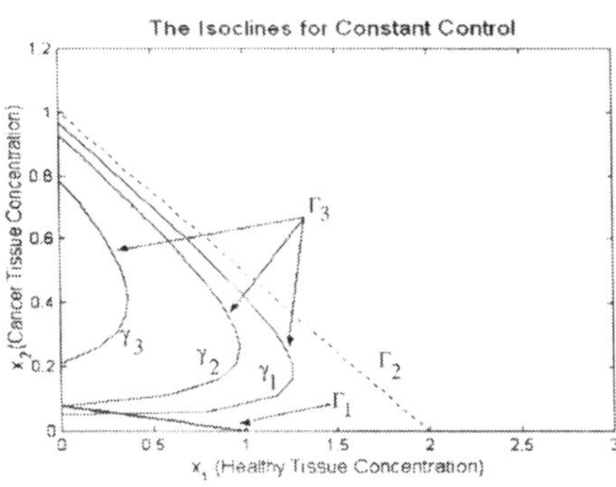

Fig. 9.2. Isoclines of (6): $a < 0$. Changes in shape of Γ_3 for different values of $\gamma : \gamma_1 < \gamma_2 < \gamma_3 < \gamma_4 < \gamma_5$

The boundary equilibria on the x_2 axis will exist if $0 = \frac{\alpha_2}{\beta_2} - \frac{\alpha_2}{\beta_2 K_2}x_2 - \frac{\gamma}{\beta_2 x_2}$ or, equivalently, $0 = \alpha_2 x_2^2 - K_2\alpha_2 x_2 + \gamma K_2$ has positive solutions. Therefore,

$$\gamma < \frac{\alpha_2 K_2}{4} \implies \text{two positive real solutions } 0 < x_2 < \frac{K_2}{2}, \ \frac{K_2}{2} < x_2 < K_2$$

$$\gamma = \frac{\alpha_2 K_2}{4} \implies \text{one positive real solution } x_2 = \frac{K_2}{2}$$

$$\gamma < \frac{\alpha_2 K_2}{4} \implies \text{no positive real solutions}.$$

$$(8)$$

To develop conditions necessary for an internal equilibrium first we solve system (7) by substituting for x_1 from the first equation into the second to obtain

$$ax_2^2 - bx_2 + \alpha_1 K_2 \gamma = 0, \tag{9}$$

where $a = \alpha_1\alpha_2 - \beta_1\beta_2 K_1 K_2$ and $b = K_2\alpha_1(\alpha_2 - K_1\beta_2)$. The solutions of this quadratic equation are given by

$$x_2 = \frac{b \pm \sqrt{b^2 - 4a\alpha_1 K_2 \gamma}}{2a}. \tag{10}$$

This x_2 defines the location of an internal equilibrium. The equilibrium from now on is labeled as $E^* = (x_1^*, x_2^*)$.

Conditions (2) $\implies b > 0$ since $\beta_2 K_1 < \alpha_2$. Variable a, however, may be positive, negative, or zero. Therefore, by conditions (6), the solution to (9) are:

$$a < 0 \implies x_2^* = \frac{b - \sqrt{b^2 - 4a\alpha_1 K_2 \gamma}}{2a} \qquad \text{is the only potential solution},$$

$$a = 0 \implies x_2^* = \frac{\gamma}{\alpha_2 - \beta_2 K_1} \qquad \text{is the only possible solution},$$

$$a > 0 \implies x_2^* = \frac{b \pm \sqrt{b^2 - 4a\alpha_1 K_2 \gamma}}{2a} \qquad \text{gives two potential solutions}.$$

$$(11)$$

There may also be a single solution when Γ_3 is tangent to Γ_1. In this case, $x_2^* = \frac{b}{2a}$ and $\gamma = \frac{b^2}{4a\alpha_1 K_2} = \frac{\alpha_1 K_2}{4a}(\alpha_2 - \beta_2 K_1)^2$, or $\gamma = \frac{a(x_2^*)^2}{K_2\alpha_1}$. In order to have a solution in the first quadrant, x_1^* should also satisfy: $0 < x_1^* < K_1$. Thus (7) $\Longrightarrow 0 < x_2^* < \frac{\alpha_1}{\beta_1}$. We obtain the following further restrictions on γ:

$$a < 0 \implies 0 < \gamma < \frac{\alpha_1\alpha_2}{\beta_1 K_2}\left(K_2 - \frac{\alpha_1}{\beta_1}\right),$$

$$a = 0 \implies 0 < \gamma < \frac{\alpha_1\alpha_2 - \alpha_1\beta_2 K_1}{\beta_1},$$

$$a > 0 \implies \begin{cases} 0 < \gamma < \dfrac{\alpha_1\alpha_2}{\beta_1 K_2}\left(K_2 - \dfrac{\alpha_1}{\beta_1}\right), & \text{(one solution)} \\[2em] \dfrac{\alpha_1\alpha_2}{\beta_1 K_2}\left(K_2 - \dfrac{\alpha_1}{\beta_1}\right) < \gamma < \dfrac{\alpha_1 K_2}{4a}(\alpha_2 - \beta_2 K_1)^2, & \text{(two solutions)}. \end{cases}$$

$$\tag{12}$$

Note that (12) must be satisfied concurrently with (2), (8) and (6) since the existence of internal solutions must guarantee the existence of solutions on the axis.

9.3.2 Stability of internal equilibria

The local stability of the internal equilibria may be determined by considering the variational matrix of system (3). Let M represent the variational matrix. Then

$$M = \begin{bmatrix} \dfrac{\partial \dot{x}_1}{\partial x_1} & \dfrac{\partial \dot{x}_1}{\partial x_2} \\[1em] \dfrac{\partial \dot{x}_2}{\partial x_1} & \dfrac{\partial \dot{x}_2}{\partial x_2} \end{bmatrix}$$

$$= \begin{bmatrix} \alpha_1\left(1 - 2\dfrac{x_1}{K_1}\right) - \beta_1 x_2 & -\beta_1 x_1 \\[1.5em] -\beta_2 x_2 & \alpha_2\left(1 - 2\dfrac{x_2}{K_2}\right) - \beta_2 x_1 \end{bmatrix}. \tag{13}$$

We would like to study the stability of the internal equilibrium, $E^* = (x_1^*, x_2^*)$. This equilibrium is found at the intersection of isoclines Γ_1 and Γ_3. Notice that when $\dot{x}_1 = 0, \beta_1 x_2 = \alpha_1\left(1 - \frac{x_1}{K_1}\right)$; and when $\dot{x}_2 = 0, \beta_2 x_1 + \frac{\gamma}{x_2} = \alpha_2\left(1 - \frac{x_2}{K_2}\right)$. Therefore, matrix (13) evaluated at $E^* = (x_1^*, x_2^*)$ is simplified to:

$$M^* = \begin{bmatrix} -\alpha_1 \dfrac{x_1^*}{K_1} & -\beta_1 x_1^* \\[1.5em] -\beta_2 x_2^* & \dfrac{\gamma}{x_2^*} - \alpha_2 \dfrac{x_2^*}{K_2} \end{bmatrix}. \tag{14}$$

The eigenvalues are the solutions of the equation

$$
\begin{aligned}
0 &= \det(\lambda I - M^*) \\
&= \lambda^2 + \lambda\left(\alpha_1\frac{x_1^*}{K_1} + \alpha_2\frac{x_2^*}{K_2} - \frac{\gamma}{x_2^*}\right) \\
&\quad + \alpha_1\frac{x_1^*}{K_1}\left(\alpha_2\frac{x_2^*}{K_2} - \frac{\gamma}{x_2^*}\right) - \beta_1\beta_2 x_1^* x_2^* .
\end{aligned}
\tag{15}
$$

If $\alpha_2\frac{x_2^*}{K_2} - \frac{\gamma}{x_2^*} < 0$, then the eigenvalues are of opposite signs and the equilibrium is a saddle point. However if $\alpha_2\frac{x_2^*}{K_2} - \frac{\gamma}{x_2^*} > 0$, then $\alpha_1\frac{x_1^*}{K_1}\left(\alpha_2\frac{x_2^*}{K_2} - \frac{\gamma}{x_2^*}\right) - \beta_1\beta_2 x_1^* x_2^*$ may be negative (a saddle point equilibrium), or positive. We simplify the expression

$$
\begin{aligned}
&\alpha_1\frac{x_1^*}{K_1}\left(\alpha_2\frac{x_2^*}{K_2} - \frac{\gamma}{x_2^*}\right) - \beta_1\beta_2 x_1^* x_2^* \\
&= \frac{x_1^*}{x_2^* K_1 K_2}[x_2^{*2}(\alpha_1\alpha_2 - \beta_1\beta_2 K_1 K_2) - \alpha_1 K_2\gamma] \\
&= \frac{x_1^*}{x_2^* K_1 K_2}[x_2^{*2} a - \alpha_1 K_2\gamma] .
\end{aligned}
$$

Since the equilibrium is located at x_2^* given by (10), we obtain the following:

$$
\begin{aligned}
&\frac{x_1^*}{x_2^* K_1 K_2}\left[\left(\frac{b \pm \sqrt{b^2 - 4a\alpha_1 K_2\gamma}}{2a}\right)^2 a - \alpha_1 K_2\gamma\right] \\
&= \frac{x_1^*}{x_2^* K_1 K_2}\left[\frac{2b^2 \pm 2b\sqrt{b^2 - 4a\alpha_1 K_2\gamma} - 4a\alpha_1 K_2\gamma}{4a} - \alpha_1 K_2\gamma\right] \\
&= \frac{x_1^*}{2ax_2^* K_1 K_2}[b^2 \pm b\sqrt{b^2 - 4a\alpha_1 K_2\gamma} - 4a\alpha_1 K_2\gamma] \\
&= \frac{x_1^*}{2ax_2^* K_1 K_2}\left[\left(\sqrt{b^2 - 4a\alpha_1 K_2\gamma}\right)^2 \pm b\sqrt{b^2 - 4a\alpha_1 K_2\gamma}\right] \\
&= \frac{x_1^*\sqrt{b^2 - 4a\alpha_1 K_2\gamma}}{2ax_2^* K_1 K_2}\left(\sqrt{b^2 - 4a\alpha_1 K_2\gamma} \pm b\right) .
\end{aligned}
$$

In the case where $a > 0$,

$$
\frac{x_1^*\sqrt{b^2 - 4a\alpha_1 K_2\gamma}}{2ax_2^* K_1 K_2}\left(\sqrt{b^2 - 4a\alpha_1 K_2\gamma} + b\right) > 0 ,
$$

$$
\frac{x_1^*\sqrt{b^2 - 4a\alpha_1 K_2\gamma}}{2ax_2^* K_1 K_2}\left(\sqrt{b^2 - 4a\alpha_1 K_2\gamma} - b\right) < 0 .
$$

These expressions correspond to $x_2^* = \frac{b+\sqrt{b^2-4a\alpha_1 K_2\gamma}}{2a}$ and to $x_2^* = \frac{b-\sqrt{b^2-4a\alpha_1 K_2\gamma}}{2a}$ respectively. In the case where $a < 0$,

$$\frac{x_1^*\sqrt{b^2-4a\alpha_1 K_2\gamma}}{2ax_2^* K_1 K_2}\left(\sqrt{b^2-4a\alpha_1 K_2\gamma}-b\right) < 0\,.$$

This expression corresponds to the only possible internal equilibrium when $a < 0$ located at $x_2^* = \frac{b-\sqrt{b^2-4a\alpha_1 K_2\gamma}}{2a}$.

Therefore, the equilibrium at $x_2^* = \frac{b-\sqrt{b^2-4a\alpha_1 K_2\gamma}}{2a}$ is a saddle point for both $a < 0$ and $a > 0$.

The equilibrium at $x_2^* = \frac{b+\sqrt{b^2-4a\alpha_1 K_2\gamma}}{2a}$ corresponds to positive $\alpha_2 \frac{x_2^*}{K_2} - \frac{\gamma}{x_2^*}$. Here

$$\alpha_1\frac{x_1^*}{K_1} + \alpha_2\frac{x_2^*}{K_2} - \frac{\gamma}{x_2^*} > 0 \Longrightarrow \mathrm{Re}(\lambda_{1,2}) < 0\,.$$

Therefore, the equilibrium at $x_2^* = \frac{b+\sqrt{b^2-4a\alpha_1 K_2\gamma}}{2a}$ is stable.

9.3.3 Conclusion

This model describes what is known, namely that the larger the value of γ, the better the control of the cancer cells. However, at the same time, the larger the γ, the greater the spillover to the healthy cells. In practical terms, a great deal of time is spent by medical researchers in finding the correct balance for radiation to control the cancer cells without doing too much damage to the normal cells.

9.4 Treatment by chemotherapy

The material from this section is based upon the Ph.D. work of Nani (1998). In the case that chemotherapy treatment is warranted, the chemotherapy agent acts like a predator on both healthy and cancer cells, by binding to them and killing them. The action of the agent on the cancer cells is desirable, but on the healthy cells is undesirable causing so-called side effects such as extreme nausea and hair loss. The object then is to design the chemotherapy agent where possible to maximize its effects on specific cancers at specific sites and to minimize the side effects.

We take as our model the system

$$\dot{x}_1 = \alpha_1 x_1\left(1 - \frac{x_1}{K_1}\right) - \beta_1 x_1 x_2 - p_1(x_1)h(y)\,, \qquad x_1(0) \geq 0$$

$$\dot{x}_2 = \alpha_2 x_2\left(1 - \frac{x_2}{K_2}\right) - \beta_2 x_1 x_2 - p_2(x_2)h(y)\,, \qquad x_2(0) \geq 0 \qquad (16)$$

$$\dot{y} = \varphi(x_1, x_2, y, t)\,, \qquad\qquad\qquad\qquad y(0) > 0$$

where $p_i(x_i)$ is the chemotherapic functional response on x_i, φ is the treatment strategy, $y(t)$ is the concentration of chemotherapy agent. $h(y)$ will be described below. All other parameters and functions are as in system (3).

Since $p_i(x_i)$ is the effect of a single chemotherapy binding site on x_i, $h(y)$ is the cumulative effects of a concentration of y binding sites. Generally $h(y)$ is nonlinear, but has the properties $h(0) = 0, h'(y) > 0$ for $y \geq 0$, there exists $0 < \overline{h} < \infty$ such that

$$\lim_{y \to \infty} h(y) = \overline{h} \tag{17}$$

(see Agur et al. 1992).

As for $p_i(x_i)$, they have the usual predator functional response properties

$$p_i(0) = 0, \quad p_i'(x_i) > 0 \quad \text{for} \quad x_i \geq 0, \tag{18}$$

(see Freedman and Waltman 1984).

$\varphi(x_1, x_2, y, t)$ will depend on the treatment strategy. We focus here on two types of treatments, namely continuous and periodic. We will discuss the continuous case in some detail, and very briefly discuss the periodic case. Details may be found in Nani (1998).

9.4.1 The continuous treatment case

In this case we take

$$\varphi(x_1, x_2, y, t) = \delta - [\gamma + \eta_1 p_1(x_1) + \eta_2 p_2(x_2)]h(y) . \tag{19}$$

Here δ is the continuous infusion of chemotherapy concentration to the affected site in question, γ is the natural washout rate, and $\eta_i, i = 1, 2$ are the binding coefficients between the chemotherapy agent and the cells.

There are four possible equilibria in this case, namely

$$E_0(0, 0, y_0), \widehat{E}_1(\widehat{x}_1, 0, \widehat{y}_1), \widehat{E}_2(0, \widehat{x}_2, \widehat{y}_2), E^*(x_1^*, y_2^*, y_3^*)$$

where y_0 is the positive solution of $h(y) = \gamma^{-1}\delta$, providing it exits.

We now show that $\widehat{E}_1$ and $\widehat{E}_2$ always exist.

Theorem 1. $\widehat{E}_i$ *always exists with* $0 < \widehat{x}_i < K_i$, $\widehat{y}_i > 0, i = 1, 2$, *provided* $\alpha_i\gamma > \delta p_i'(0)$ *and* $\overline{h} > \delta\gamma^{-1}$.

Proof. We prove this for the case $i = 1$. The case $i = 2$ follows analogously. $\widehat{x}_1$ and $\widehat{y}_1$ satisfy the system

$$\alpha_1\widehat{x}_1\left(1 - \frac{\widehat{x}_1}{K_1}\right) - p_1(\widehat{x}_1)h(\widehat{y}) = 0 \tag{20}$$

$$\delta - [\gamma + \eta_1 p_1(\widehat{x}_1)]h(\widehat{y}) = 0 .$$

Substituting

$$h(\widehat{y}) = \frac{\delta}{\gamma + \eta_1 p_1(\widehat{x}_1)} \tag{21}$$

into the first equation of (20) and writing $p_1(\widehat{x}_1) = \widehat{x}_1\widetilde{p}_1(\widehat{x}_1)$ (since $p_1(0) = 0$ and $p_1'(0)$ exists), we get that for $\widehat{x}_1 > 0$,

$$\alpha_1\left(1 - \frac{\widehat{x}_1}{K_1}\right)(\gamma + \eta_1\widehat{x}_1\widetilde{p}_1(\widehat{x}_1)) = \delta\widetilde{p}_1(\widehat{x}_1). \tag{22}$$

Note that $\widetilde{p}_1(0) = p_1'(0) > 0$. Writing (22) as $F_1(\widehat{x}_1) = G_1(\widehat{x}_1)$, we easily see that $F_1(0) = \alpha_1\gamma > 0, F_1(K_1) = 0, G_1(0) = \delta p_1'(0), G_1(K_1) = \delta\widetilde{p}_1(K_1) > 0$. Since by hypothesis $F_1(0) > G_1(0)$ and $F_1(K_1) < G_1(K_1)$, there exists a $0 < \widehat{x}_1 < K_1$ such that (22) holds. Then from (21), $h(\widehat{y}) > 0$ exists and therefore $\widehat{y} > 0$ exists.

To check whether E^* exists, one must solve the full algebraic system, writing $p_i(x_i) = x_i\widetilde{p}_i(x_i), i = 1, 2$,

$$\alpha_1\left(1 - \frac{x_1}{K_1}\right) - \beta_1 x_2 - \widetilde{p}_1(x_1)h(y) = 0$$

$$\alpha_2\left(1 - \frac{x_2}{K_2}\right) - \beta_2 x_1 - \widetilde{p}_2(x_2)h(y) = 0 \tag{23}$$

$$\delta - [\gamma + \eta_1 x_1\widetilde{p}_1(x_1) + \eta_2 x_2\widetilde{p}_2(x_2)]h(y) = 0 .$$

Substituting

$$h(y) = \frac{\delta}{\gamma + \eta_1 x_1\widetilde{p}_1(x_1) + \eta_2 x_2\widetilde{p}_2(x_2)} \tag{24}$$

into the first two equations of (23) gives the algebraic system

$$\left[\alpha_1\left(1 - \frac{x_1}{K_1}\right) - \beta_1 x_2\right][\gamma + \eta_1 x_1\widetilde{p}_1(x_1) + \eta x_2\widetilde{p}_2(x_2)] = \delta\widetilde{p}_1(x_1)$$

$$\left[\alpha_2\left(1 - \frac{x_2}{K_2}\right) - \beta_2 x_1\right][\gamma + \eta_1 x_1\widetilde{p}_1(x_1) + \eta_2 x_2\widetilde{p}_2(x_2)] = \delta\widetilde{p}_2(x_2) . \tag{25}$$

As before, if $x_1^*, x_2^* > 0$ exists, then from (24) so does $y^* > 0$.

It is extremely difficult to see whether or not system (25) has a positive solution. Hence we take a different approach to obtain criteria for the existence of E^*, namely persistence theory. In order to do so, we will need the variational matrices about $E_0, \widehat{E}_1$, and $\widehat{E}_2$.

The general variational matrix about an equilibrium $(\overline{x}_1, \overline{x}_2, \overline{y})$ is given by

$$\overline{M} = \begin{bmatrix} \alpha_1\left(1 - \frac{2\overline{x}_1}{K_1}\right) - \beta_1\overline{x}_2 & -\beta_1\overline{x}_1 & -p_1(\overline{x}_1)h'(\overline{y}) \\ \quad -p_1'(\overline{x}_1)h(\overline{y}) & & \\ -\beta_2\overline{x}_2 & \alpha_2\left(1 - \frac{2\overline{x}_2}{K_2}\right) - \beta_2\overline{x}_1 & -p_2(\overline{x}_2)h'(\overline{y}) \\ & \quad -p_2'(\overline{x}_2)h(\overline{y}) & \\ -\eta_1 p_1'(\overline{x}_1)h(\overline{y}) & -\eta_2 p_2'(\overline{x}_2)h(\overline{y}) & -[\gamma + \eta_1 p_1(\overline{x}_1) \\ & & +\eta_2 p_2(\overline{x}_2)]h'(\overline{y}) \end{bmatrix} .$$

This implies, after some simplifications

$$M_0 = \begin{bmatrix} \alpha_1 - p_1'(0)h(y_0) & 0 & 0 \\ 0 & \alpha_2 - p_2'(0)h(y_0) & 0 \\ -\eta_1 p_1'(0)h(y_0) & -\eta_2 p_2'(0)h(y_0) & -\gamma h'(y_0) \end{bmatrix}$$

$$\widehat{M}_1 = \begin{bmatrix} -\frac{\alpha_1 \widehat{x}_1}{K_1} + \{\widetilde{p}_1(\widehat{x}_1 \\ -p_1'(\widehat{x}_1)\}h(\widehat{y}_1) & -\beta_1 \widehat{x}_1 & -p_1(\widehat{x}_1)h'(\widehat{y}_1) \\ 0 & \alpha_2 - \beta_2 \widehat{x}_1 - p_2'(0)h(\widehat{y}_1) & 0 \\ -\eta_1 p_1'(\widehat{x}_1)h(\widehat{y}_1) & -\eta_2 p_2'(0)h(\widehat{y}_1) & -[\gamma + \eta_1 p_1(\widehat{x}_1)]h'(\widehat{y}_1) \end{bmatrix}$$

$$\widehat{M}_2 = \begin{bmatrix} \alpha_1 - \beta_1 \widehat{x}_2 - p_1'(0)h(\widehat{y}_2) & 0 & 0 \\ -\beta_2 \widehat{x}_2 & -\frac{\alpha_2 \widehat{x}_2}{K_2} + \{\widetilde{p}_2(\widehat{x}_2) \\ -p_2'(\widehat{x}_2)\}h(\widehat{y}_2) & -p_2(\widehat{x}_2)h'(\widehat{y}_2) \\ -\eta_1 p_1(0)h(\widehat{y}_2) & -\eta_2 p_2'(\widehat{x}_2)h(\widehat{y}_2) & -[\gamma + \eta_2 p_2(\widehat{x}_2)]h'(\widehat{y}_2) \end{bmatrix} .$$

First we examine M_0. The eigenvalues of M_0 are given by

$$\alpha_1 - p_1'(0)h(y_0) , \quad \alpha_2 - p_2'(0)h(y_0) \quad \text{and} \quad -\gamma h'(y_0) .$$

From this, E_0 is clearly locally stable in the y direction and is locally stable or unstable in the x_i direction according to whether $\alpha_i - p_i'(0)h(y_0)$ is negative or positive.

The important concern is with $\alpha_2 - p_2'(0)h(y_0)$, for if this expression is negative, then cancer can be eradicated if caught in time. However, at the same time we would want $\alpha_1 - p_1'(0)h(y_0) > 0$ so that the healthy cells survive.

Finally for persistence to hold according to techniques developed in Freedman and Waltman (1984), we would require E_0 to be unstable, and for $\widehat{E}_i$ to be unstable locally in the j direction, $i, j = 1, 2, j \neq i$. Hence the criteria for persistence are as follows:

$$\begin{aligned} \alpha_1 - \beta_1 \widehat{x}_2 - p_1'(0)h(\widehat{y}_2) &> 0 , \\ \alpha_2 - \beta_2 \widehat{x}_1 - p_2'(0)h(\widehat{y}_1) &> 0 , \end{aligned} \tag{26}$$

and one of

$$\alpha_i - p_i'(0)h(y_0) > 0 , \quad i = 1, 2 .$$

Finally from results given in Butler et al. (1986), if (26) holds, then E^* exists.

9.4.2 Periodic treatment

In actual practice, a form of periodic treatment is employed. Typically, the cancer patient is given a fixed number of doses over a fixed period of time at regular intervals. This may be approximated by a periodic step function.

In general, we let

$$\varphi(x_1, x_2, y, t) = f(t) - [\delta + \eta_1 p_1(x_1) + \eta_2 p_2(x_2)]h(y) , \qquad (27)$$

where $f(t) \geq 0$ and $f(t + \omega) = f(t)$. With this form of $\varphi(x_1, x_2, y, t)$ as given by (27), there can be no interior equilibrium. Hence if cancer cannot be forced to extinction (which is the usual case), criteria need to be developed for there to exist a positive periodic solution to system (16) with low values of x_2. I will now briefly describe how to develop these criteria, but due to their complexity, will not state them here.

First note that by Massera's theorem (see Pliss (1966)) there is a positive periodic solution on the y-axis. Then using some standard bifurcation theory, one obtains criteria for a positive periodic solution in the $x_1 - y$ plane.

Now comes the tricky part. The idea is to develop criteria for this solution to bifurcate away from the plane into the positive $x_1 - x_2 - y$ space. One way of doing this is to use critical cases of the implicit function theorem (see Nani (1998)) and so obtain the required criteria.

9.4.3 Conclusion

Under appropriate circumstances, a periodic application of chemotherapy may force a periodic behaviour in the interactions between healthy and cancer cells and the chemotherapy agent. Again, this would be most likely if the cancer is detected at an early stage.

9.5 Treatment by immunotherapy

The material in this section is based on work done in Nani and Freedman (2000).

When cancer cells proliferate to a detectable threshold number at a given site, the body's own natural immune system is triggered into a search-and-destroy mode. Unfortunately, the process of natural immune attack against immunogenic cancer is not always sustainable nor eventually successful and can always be terminated or downgraded due to various reasons, including insufficient lymphocytes, evasion by cancer cells or release of inhibitory substances by the cancer cells (Toledo-Pereya 1988), and for these reasons, the natural immune system cannot provide a therapeutically successful anti cancer attack.

This can be overcome to some extent by clinically extracting lymphocytes from the body, incubating these so called LAK cells outside the body for at least 48 hours, and then reintroducing them into the body.

This leads to the following model consisting of four ODEs:

$$\dot{x}_1 = \alpha_1 x_1 \left(1 - \frac{x_1}{K_1}\right) - \beta_1 x_1 x_2 \,, \qquad\qquad x_1(0) \geq 0$$

$$\dot{x}_2 = \alpha_2 x_2 \left(1 - \frac{x_2}{K_2}\right) - \beta_1 x_1 x_2 - h(x_3, w) \,, \qquad x_2(0) \geq 0 \qquad (28)$$

$$\dot{w} = Q_1 - \gamma_1 e_1(w) + f(w, z) - \delta h(x_2, w) \,, \qquad w(0) \geq 0$$

$$\dot{z} = Q_2 - \gamma_2 e_2(z) - \eta f(w, z) \,, \qquad\qquad z(0) \geq 0 \,.$$

Here $w(t)$ is the concentration of lymphocytes $z(t)$ is the concentration of LAC cells, $f(w, z)$ is the rate of lymphocyte proliferation due to the influence of LAC cells, $h(x_2, z)$ is the rate of cancer destruction by lymphocytes and Q_i are the respective rates of infusion of lymphocytes and LAC cells into the body. $\gamma_1 e_1(w)$ and $\gamma_2 e_2(z)$ are the natural death or washout rates of the lymphocytes and LAC cells respectively. δ is the proportionate combination of lymphocytes with cancer cells, and η is the proportionate influence of the lymphocytes on LAC cells. It is shown in Nani and Freedman (2000) that solutions of system (28) enter into a bounded invariant region and that the system is dissipative.

There arc four possible equilibria for system (28) of the form

$$E_0(0, 0, \overset{\circ}{w}, \overset{\circ}{z}) \,, \quad E_1(\overline{x}_1, 0, \overline{w}, \overline{z}) \,, \quad E_2(0, \widehat{x}_2, \widehat{w}, \widehat{z})$$

and

$$E_3(x_1^*, x_2^*, w^*, z^*) \,.$$

The equilibrium of interest is E_1, for if E_1 is locally stable in the x_2 direction, then cancer could be eradicated if caught early enough. The variational matrix of system (28) about E_1, assuming it exists is

$$\begin{bmatrix} -\alpha_1 & -\beta_1 K_1 & 0 & 0 \\ 0 & \begin{matrix}(\alpha_2 - \beta_2 K_1 \\ -h_{x_2}(0, \overline{w}))\end{matrix} & -h_w(0, \overline{w}) & 0 \\ 0 & -\delta h_{x_2}(0, \overline{w}) & \begin{matrix}-\gamma_1 e_1'(\overline{w}) + f_w(\overline{w}, \overline{z}) \\ -\delta h_w(0, \overline{w})\end{matrix} & f_z(\overline{w}, \overline{z}) \\ 0 & & -\eta f_w(\overline{w}, \overline{z}) & -\gamma_2 e_2'(\overline{z}) - \eta f_z(\overline{w}, \overline{z}) \end{bmatrix} \,.$$

Then the local stability in the x_2 direction is given by $g(\overline{w}) = \alpha_2 - \beta_2 K_1 - h_{x_2}(0, \overline{w})$ assuming $h_w(0, w) = 0$ given the definition of $h(x_2, w)$. Hence if $g(\overline{w}) < 0$, cancer can be eradicated if caught early enough.

System (28) is analyzed in detail in Nani and Freedman (2000).

9.6 Metastasis

Metastasis means that the cancer has spread from one site to another. Usually the metastasis occurs one way only. It is very often the case that the cancer at the second site is much more deadly than at the first site. The material from this section is taken from Pinho et al. (2002).

We consider cancer at two sites treated by chemotherapy. This requires a system of six ODE's. We let $x_1(t)$ and $x_2(t)$ be the concentration of healthy and cancer cells respectively at the primary site and $u_1(t)$ and $u_2(t)$ be the concentration of healthy cells and cancer cells respectively at the secondary site. We further let $y(t)$ and $z(t)$ be the concentration of chemotherapy agent at the primary and secondary sites respectively. Thus our model becomes

$$\dot{x}_1(t) = \alpha_1 x_1(t)\left(1 - \frac{x_1(t)}{K_1}\right) - \beta_1 x_1(t)x_2(t) - \frac{p_1 x_1(t)y(t)}{a_1 + x_1(t)},$$

$$x_1(0) \geq 0$$

$$\dot{x}_2(t) = \alpha_2 x_2(t)\left(1 - \frac{x_2(t)}{K_2}\right) - \beta_2 x_1(t)x_2(t) - \frac{p_2 x_2(t)y(t)}{a_2 + x_2(t)} - \theta x_2(t),$$

$$x_2(0) \geq 0$$

$$\dot{y}(t) = \Delta - \left[\xi + \frac{c_1 x_1(t)}{a_1 + x_1(t)} + \frac{c_2 x_2(t)}{a_2 + x_2(t)}\right]y(t),$$

$$y(0) \geq 0$$

$$\dot{u}_1(t) = \gamma_1 u_1(t)\left(1 - \frac{u_1(t)}{L_1}\right) - \delta_1 u_1(t)u_2(t) - \frac{s_1 u_1(t)z(t)}{b_1 + u_1(t)},$$

$$u_1(0) \geq 0$$

$$\dot{u}_2(t) = \gamma_2 u_2(t)\left(1 - \frac{u_2(t)}{L_2}\right) - \delta_2 u_1(t)u_2(t) - \frac{s_2 u_2(t)z(t)}{b_2 + u_2(t)} + \varepsilon\theta x_2(t - \tau),$$

$$u_2(0) \geq 0$$

$$\dot{z}(t) = \Phi - \left[\eta + \frac{d_1 u_1(t)}{b_1 + u_1(t)} + \frac{d_2 u_2(t)}{b_2 + u_2(t)}\right]z(t),$$

$$z(0) \geq 0,$$

$$(29)$$

where we have chosen specific functional responses for simplicity and all other constants have similar interpretations as before.

Here the new feature in this model is the introduced delay term in the fifth equation, $\varepsilon\theta x_2(t - \tau)$, which represents the fact that it takes time τ for the cancer growth to be triggered at the secondary site. Here θ is the proportion of cancer cells from the first site that are activated at the secondary site, a_i are the respective Michaelis-Menton growth constants for x_i, and b_i are similar for u_i. Note that system (29) simulates the continuous treatment case.

System (29) has nine possible equilibria of the form

$$F_0(0,0,\xi^{-1}\Delta,0,0,\eta^{-1}\Phi)\,, \qquad F_1(\widehat{x}_1,0,\widehat{y},0,0,\eta^{-1}\Phi)$$
$$F_2(0,0,\xi^{-1}\Delta,\widehat{u}_1,0,\widehat{z})\,, \qquad F_3(\widehat{x}_1,0,\widehat{y},\widehat{u}_1,0,\widehat{z})\,,$$
$$F_4(\widehat{x}_1,0,\widehat{y},0,\overline{u}_2,\overline{z})\,, \qquad F_5(0,\check{x}_2,\check{y},0,\check{u}_2,\check{z})$$
$$F_6(x_1^*,x_2^*,y^*,0,u_2^{\#},z^{\#})\,, \qquad F_7(0,\check{x}_2,\check{y},u_1^{\dagger},u_2^{\dagger},z^{\dagger})\,,$$
$$F_8(x_1^*,x_2^*,y^*,u_1^*,u_2^*,z^*)\,.$$

These are extremely difficult to analyze analytically. Here we will give some numerical results. A more detailed analysis can be found in Pinho et al. (2002).

The following three figures indicate a variety of behaviours of solutions, depending on parameters and initial conditions.

In Fig. 9.3, we see that at the primary site, cancer is eradicated, but at the secondary site after time τ, the cancer takes over and drives the healthy cells to extinction. Unfortunately, this is all too often the case.

In Fig. 9.4, the behaviour at the primary site is the same as in Fig. 9.3, but at the secondary site, wild chaotic oscillations occur. This unpredictability makes it extremely difficult to prescribe treatment. This corresponds to cases where cancer seems to go in and out of remission until the body succumbs.

Finally Fig. 9.5 shows that for certain cancers and chemotherapies, the cancer can be controlled at both sites.

(a)   (b)

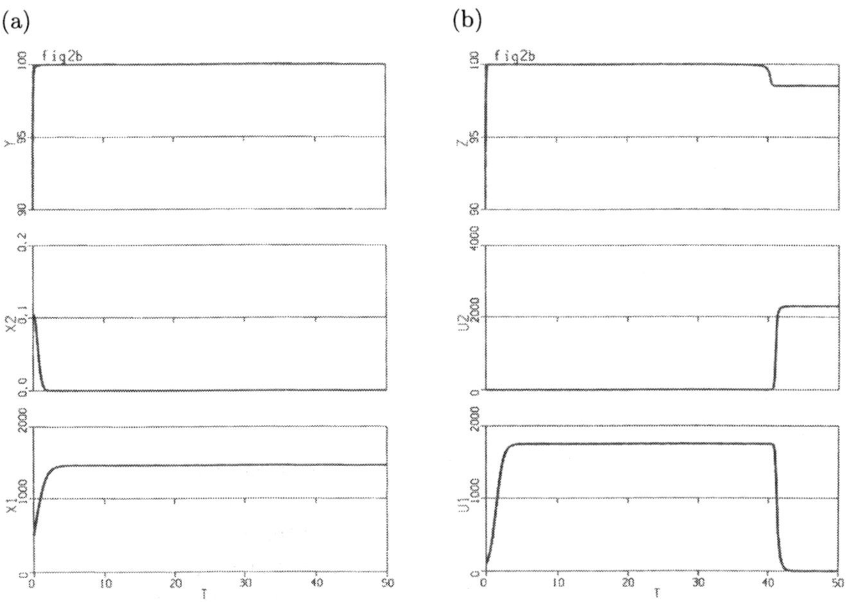

Fig. 9.3. a Cancer eradicated at primary site. **b** Cancer outcompetes normal cells at secondary site

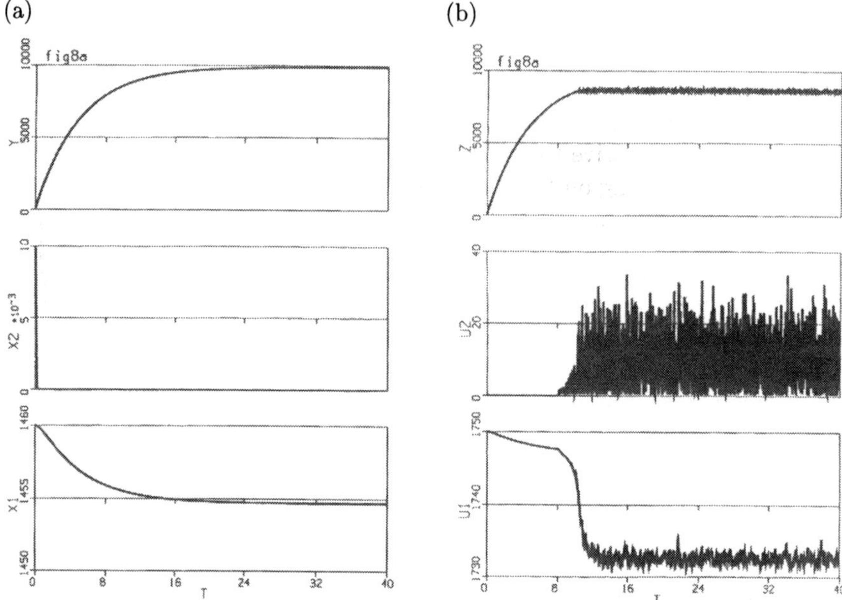

Fig. 9.4. a Cancer eradicated at primary site. **b** Chaotic behavior at secondary site

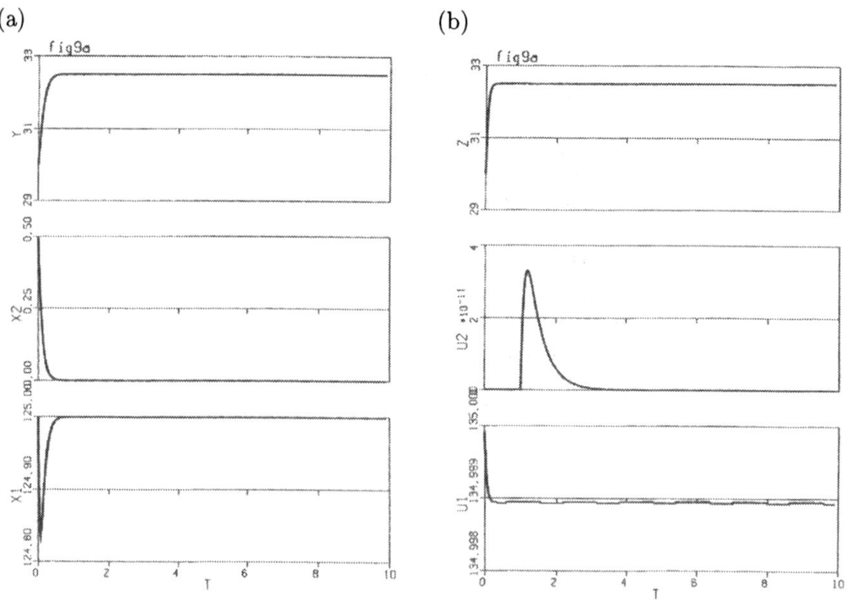

Fig. 9.5. Cancer eradicated at both primary and secondary sites

9.7 Discussion

In this paper we have briefly described various models of cancer treatment by radiotherapy, chemotherapy and immunotherapy. In all cases, we have shown that it is possible to drive the cancer extinct provided that it is caught early enough, and depending on the type of cancer.

However, we note that there are certain types of cancers, such as leukemia, for which these models do not apply. It is the purpose of future investigations to develop more robust models which do apply to other cancers.

Acknowledgement. The author wishes to thank an anonymous referee for a careful reading of the manuscript.

References

1. Agur, Z., R. Arnon and B. Schector (1992), Effect of dosing interval on myelotoxicity and survival in mice treated by cytarabine, Eur. J. Cancer **28**, 1085–1090.
2. Belostotski, G. (2004), A Control Theory Model for Cancer Treatment by Radiotherapy, M.Sc. Thesis, University of Alberta.
3. Butler, G.J., H.I. Freedman and P. Waltman (1986), Uniformly persistent systems, Proc. Amer. Math. Soc. **96**, 425–430.
4. Freedman, H.I. and P. Waltman (1984), Persistence in models of three interacting predator-prey populations, Math. Biosci. **68**, 213–231.
5. Freedman, H.I. (1980), Deterministic Mathematical Models in Population Ecology, Marcel Dekker, New York.
6. Horn, M.A. and G. Webb (2004), Discrete and continuous dynamical systems **4**, 1–348, special issue on Mathematical Models in Cancer.
7. Nani, F. and H.I. Freedman (2000), A mathematical model of cancer treatment of immunotherapy, Math. Biosci. **163**, 159–199.
8. Nani, F. (1998), Mathematical Models of Chemotherapy and Immunotherapy, Ph.D. Thesis, University of Alberta.
9. Pinho, S.T.R., H.I. Freedman and F. Nani (2002), A chemotherapy model for the treatment of cancer with metastasis, Math. Comput. Model **36**, 773–803.
10. Pliss, V.A. (1966), Nonlocal Problems of the Theory of Oscillations, Academic Press, New York.
11. Toledo-Pereya, L.H. (1988), Immunology Essentials for Surgical Practice, PSG, Littlestone, MA.

Index

Printing: Krips bv, Meppel
Binding: Stürtz, Würzburg

Breinigsville, PA USA
17 May 2010
238034BV00005B/17/P